Ehrenstein / Riedel / Trawiel
Thermal Analysis of Plastics

About the Cover Illustration:

POM-C

10-μm thin section
Polarized transmitted light with lambda plate
Scale: 1 to 100

Gottfried W. Ehrenstein
Gabriela Riedel
Pia Trawiel

Thermal Analysis of Plastics

Theory and Practice

HANSER

Hanser Publishers, Munich • Hanser Gardner Publications, Cincinnati

The Authors:
Prof. Dr.-Ing. Dr. h.c. Gottfried W. Ehrenstein, Dipl.-Ing. (FH) Gabriela Riedel, Pia Trawiel, Lehrstuhl für Kunststofftechnik Universität Erlangen-Nürnberg, Am Weichselgarten 9, D-91058 Erlangen-Tennenlohe, Germany, Tel.: +49 (91 31) 85 – 2 97 00, Fax.: +49 (91 31) 85 – 02 97 09, E-mail: ehrenstein@lkt.uni-erlangen.de, Internet: http://www.lkt.uni-erlangen.de

Distributed in the USA and in Canada by
Hanser Gardner Publications, Inc.
6915 Valley Avenue, Cincinnati, Ohio 45244-3029, USA
Fax: (513) 527-8801
Phone: (513) 527-8977 or 1-800-950-8977
Internet: http://www.hansergardner.com

Distributed in all other countries by
Carl Hanser Verlag
Postfach 86 04 20, 81631 München, Germany
Fax: +49 (89) 98 48 09
Internet: http://www.hanser.de

The use of general descriptive names, trademarks, etc., in this publication, even if the former are not especially identified, is not to be taken as a sign that such names, as understood by the Trade Marks and Merchandise Marks Act, may accordingly be used freely by anyone.
While the advice and information in this book are believed to be true and accurate at the date of going to press, neither the authors nor the editors nor the publisher can accept any legal responsibility for any errors or omissions that may be made. The publisher makes no warranty, express or implied, with respect to the material contained herein.

Library of Congress Cataloging-in-Publication Data
Ehrenstein, G. W.
[Praxis der thermischen Analyse von Kunststoffen. English]
Thermal analysis of plastics / Gottfried W. Ehrenstein, Gabriela Riedel, Pia Trawiel.
p. cm.
Includes bibliographical references and index.
ISBN 1-56990-362-X
1.Thermal analysis. I. Riedel, Gabriela. II. Trawiel, Pia. III.
Title.
QD79.T38E3713 2004
543'.26--dc22

                                    2004017191

Bibliografische Information Der Deutschen Bibliothek
Die Deutsche Bibliothek verzeichnet diese Publikation in der Deutschen Nationalbibliografie;
detaillierte bibliografische Daten sind im Internet über <http://dnb.ddb.de> abrufbar.
ISBN 3-446-22673-7

© Carl Hanser Verlag, Munich 2004
Production Management: Oswald Immel
Typeset by Trawiel/Riedel GbR, Germany
Coverconcept: Marc Müller-Bremer, Rebranding, München, Germany
Coverdesign: MCP • Susanne Kraus GbR, Holzkirchen, Germany
Printed and bound by Kösel, Krugzell, Germany

Preface

Over the years, we have found thermoanalytical techniques to be powerful, highly versatile tools for conducting research projects in the fields of thermoplastic, thermoset and elastomer processing, practical development and testing, tribology and bonding technology, electronics and material composites, materials testing and materials failure testing. By virtue of the vast number of experiments we conducted over this period and the fact that we employed instruments from different manufacturers as well as complementary techniques, we have gained major insights into both the advantages and the daily tribulations associated with thermoanalytical methods. These span all aspects of thermal analysis: the techniques themselves, the preparation of samples, performance of experiments and interpretation of results as well as the practicability of the techniques in real life and the scientific explanations of the relationships involved.

Such work calls for seamless, across-the-board cooperation between production technicians, mechanical engineers, materials scientists, chemical engineers, physicists and chemists, scientists, project engineers and laboratory staff, whether they be specialists or non-specialists. For these people, thermal analysis is first and foremost a working tool and not an end in itself. And it is for these people and the many others facing similar problems that we have written this book – to serve firstly as an aid to practical work, to handling instruments, preparing samples, estimating settings, interpreting results, assessing accuracy and reproducibility, warning against over-estimation, and critically assessing the interpretations. The second goal is to acquaint them with the many advantages and vast potential that thermoanalytical techniques have to offer.

There will undoubtedly be inadvertent omissions in the book – as well as room for improvement. Your feedback is always welcome. We would like to express our appreciation to the many friends, colleagues and assistants who offered us helpful comments and encouragement, to whom we are indebted for their suggestions and advice and without whom this book would never have contained the wealth of information that it does. Our sincere gratitude is extended to Prof. Dr. Tim Osswald, Prof. Dr. Jozef Varga, Prof. Helmut Vogel, Dr. Eva Bittmann, Prof. Dr. Erich Kramer, Dr. Klaus Könnecke, Dr. Jens Rieger, Dr. Herbert Stutz, Dr. Ingolf Hennig, and from the Lehrstuhl für Kunststofftechnik, to Dr. Sonja Pongratz, Dr. Johannes Wolfrum, Prof. Dr. Michael Schemme and Juditha Hudi – all of whom assisted, advised and positively criticized our work in some form or another. We would also like to thank Raymond Brown for the translation and Dr. Duncan M. Price for the critical review of the work. Last but by no means least, we would like to thank the "Grand Old Lady of Thermal Analysis", Prof. Edith Turi, for encouraging us unremittingly and endorsing our approach to the subject matter.

Gottfried W. Ehrenstein
Gabriela Riedel
Pia Trawiel

Table of Contents

Standards for Thermal Analysis

Differential Scanning Calorimerty (DSC)

ASTM D 3417 (1999)
Standard Test Method for Enthalpies of Fusion and Crastallization of Polymers by Differential Scanning Calorimetry (DSC)

ASTM D 3418 (1999)
Standard Test Method for Transition Temperatures by Differential Scanning Calorimetry

ASTM D 4591 (2001)
Standard Test Method of Determining Temperatures and Heats of Transitions of Fluoropolymers by Differential Scanning Calorimetry

ASTM E 793 (2001)
Standard Test Method for Enthalpies of Fusion and Crystallization by Differential Scanning Calorimetry

ASTM E 794 (2001)
Standard Test Method for Melting Temperatures and Crystallization Temperatures by Thermal Analysis

ASTM E 967 (2003)
Standard Practice for Temperature Calibration of Differential Scanning Calorimeters and Differential Thermal Analyzers

ASTM E 968 (2002)
Standard Practice for Heat Flow Calibration of Differential Scanning Calorimeters

ASTM E 1269 (2001)
Standard Test Method of Determining Specific Heat Capacity by Differential Scanning Calorimetry

ASTM E 1356 (2003)
Standard Test Method for Assignment of the Glass Transition Temperatures by Differential Scanning Calorimetry

ASTM 2069 (2000)
Standard Test Method for Temperature Calibration on Cooling of Differential Scanning Calorimeters

ASTM E 2070 (2000)
Standard Test Method for Kinetic Parameters by Differential Scanning Calorimetry Using Isothermal Methods

ISO 11357-1 (1997)
Plastics – Differential Scanning Calorimetry (DSC) – Part 1: General Principles

ISO 11357-2 (1999)
Plastics – Differential Scanning Calorimetry (DSC) – Part 2: Determination of Glass Transition Temperature

ISO 11357-3 (1999)
Plastics – Differential Scanning Calorimetry (DSC) – Part 3: Determination of Temperature and Enthalpy of Melting and Crystallization / Note: To be amended by ISO 11357-3 DAM 1 (2004)

ISO/DIS 11357-4 (2003)
Plastics – Differential Scanning Calorimetry (DSC) – Part 4: Determination of Specific Heat Capacity

ISO 11357-5 (1999)
Plastics – Differential Scanning Calorimetry (DSC) – Part 5: Determination of Characteristic Reaction Curve Temperatures and Times, Enthalpie of Reaction and Degree of Conversion / Note: EQV DIN 53765 (1994)

ISO 11357-7 (2002)
Plastics – Differential Scanning Calorimetry (DSC) – Part 7: Determination of Crystallization Kinetics

ISO/DIS 11357-8 (2001)
Plastics – Differential Scanning Calorimetry (DSC) – Part 8: Determination of Amount of Absorbed Water

ISO 11409 (1993)
Plastics – Phenolic Resins – Determination of Heats and Temperatures of Reactions by Differential Scanning Calorimetry

prEN 6041 (1995)
Aerospace Series – Non-metallic Materials – Test Method; Analysis of Non-metallic (uncured) by Differential Scanning Calorimetry (DSC)

prEN 6064 (1995)
Aerospace Series – Non-metallic Materials – Analysis of Non-metallic Materials (cured) for the Determination of the Extend of Cure by Differential Scanning Calorimetry (DSC)

DIN 51 005 (1999)
Thermal Analysis (TA) – Terms / Note: Intended as Replacement for DIN 51 005 (1993)

DIN 51 007 (1994)
Thermal Analysis – Differential Thermal Analysis; Principles

DIN 53 765 (1994)
Testing of Plastics an Elastomers – Thermal Analysis; DSC-Method / Note: EQV ISO
11357-5 (1999)

DIN 65 467 (1999)
Aerospace – Testing of Thermosetting Resin Systems with and without Reinforcement –
DSC-Method

Oxidative Induction Time/Temperature (OIT)

ASTM D 3895 (2003)
Standard Test Method for Oxidative Induction Time of Polyolefins by Differential
Scanning Calorimetry

ASTM D 5885 (1997)
Standard Test Method for Oxidative Induction Time of Polyolefin Geosynthetics by
High-Pressure Differential Scanning Calorimetry

ASTM E 2009 (2002)
Standard Test Method for Temperature Calibration on Cooling of Differential Scanning
Calorimeters

ISO 11357-6 (2002)
Plastics – Differential Scanning Calorimetry (DSC) – Part 6: Determination of Oxidation
Induction Time

DS 2131.2 (1982)
Dansk Standard, Pipes, Fittings and Joints of Polyethylene
Type PEM and PEH for Buried Gas Pipelines

Thermogravimetry (TG)

ASTM D 6370 (1999)
Standard Test Method for Rubber-Compositional Analysis by Thermogravimetry (TGA) /
Note: Reapproved 2003

ASTM E 1131 (2003)
Standard Test Method for Compositional Analysis by Thermogravimetry

ASTM E 1582 (2000)
Standard Test Method for Rubber-Compositional Analysis by Thermogravimetry (TGA) /
Note: Reapproved 2003

ASTM E 1641 (1999)
Standard Test Mehod for Decomposition Kinetics by Thermogravimetry

ASTM E 1868 (2002)
Standard Test Method for Loss-On-Drying by Thermogravimetry

ASTM E 2008 (1999)
Standard Test Method for Volatility Rate by Thermogravimetry

ISO 9924-1 (2000)
Rubber and Rubber Products – Determination of the Composition of Vulcanizates and Uncured Compounds by Thermogravimetry – Part 1: Butadiene, Ethylene-Propylene Copolymer and Terpolymer, Isobutene-Isoprene, Isoprene and Styrene-Butadiene Rubbers

ISO 9924-2 (2000)
Rubber and Rubber Products – Determination of the Composition of Vulcanizates and Uncured Compounds by Thermogravimetry – Part 2: Acrylonitrile-Butadiene and Halobutyl Rubbers

ISO 11358 (1997)
Plastics – Thermogravimetry (TG) of Polymers – Gerneral Principles

ISO/DIS 21870 (2003)
Rubber Compounding Ingredients – Carbon Black – Determination of High-Temperature Loss on Heating by Thermogravimetry

prEN 1878 (1995)
Products and Systems for the Protection and Repair of Concrete Structures – Test Methods – Reactive Functions Related to Epoxy Resins – Thermogravimetry of Polymers – Temperature Scanning Method

DIN 51 006 (2000)
Thermal Analysis (TA), Thermogravimetry (TG) – Principles

Thermomechanical Analysis (TMA)

ASTM E 831 (2003)
Standard Test Method for Linear Thermal Expansion of Solid Materials by Thermomechanical Analysis

ASTM E 1363 (2003)
Standard Test Method for Temperature Calibration of Thermomechanical Anaylzers

ASTM E 1545 (2000)
Standard Test Method for Assignment of the Glass Transition Temperature by Thermomechanical Analysis

ASTM E 1824 (2002)
Standard Test Method for Assignment of a Glass Transition Temperature Using Thermomechanical Analysis Under Tension

ASTM E 2113 (2002)
Standard Test Method for Length Change Calibration of Thermomechanical Analyzers

ISO 11359-1 (1999)
Plastics – Thermomechanical Analysis (TMA) – Part 1: General Principles

ISO 11359-2 (1999)
Plastics – Thermomechanical Analysis (TMA) – Part 2: Determination of Coefficient of Linear Thermal Expansion and Glass Transition Temperature

ISO 11359-3 (2002)
Plastics – Thermomechanical Analysis (TMA) – Part 3: Determination of Penetration Temperature

DIN 53 752 (1980)
Testing of Plastics – Determination of the Coefficient of Linear Thermal Expansion

Dynamic Mechanical Analysis (DMA)

ASTM D 4065 (2001)
Standard Practice for Plastics – Dynamic Mechanical Properties – Determination and Report of Procedures

ASTM D 4092 (2001)
Standard Terminology – Plastics – Dynamic Mechanical Properties

ASTM D 4440 (2001)
Standard Test Method for Plastics – Dynamic Mechanical Properties – Melt Rheology

ASTM D 4473 (2003)
Standard Test Method for Plastics – Dynamic Mechanical Properties – Cure Behavior

ASTM D 5023 (2001)
Standard Test Method for Measuring the Dynamic Mechanical Properties – In Flexure (Three-Point-Bending)

ASTM D 5024 (2001)
Standard Test Method for Plastics – Dynamic Mechanical Properties – In Compression

ASTM D 5026 (2001)
Standard Test Method for Plastics – Dynamic Mechanical Properties – In Tension

ASTM D 5279 (2001)
Standard Test Method for Plastics: Dynamic Mechanical Properties: In Torsion

ASTM D 5418 (2001)
Standard Test Method for Plastics – Dynamic Mechanical Properties – In Flexure (Dual Cantilever Beam)

ASTM D 6048 (2002)
Standard Practice for Stress Relaxation Testing of Raw Rubber, Unvulcanized Rubber
Compounds, and Thermoplastic Elastomers

ASTM E 1640 (1999)
Standard Test Method for Assignment of the Glass Transition Temperature by Dynamic
Mechanical Analysis

ASTM E 1867 (2001)
Standard Test Method for Temperature Calibration of Dynamic Mechanical Analyzers

ASTM E 2254 (2003)
Standard Test Method for Storage Modulus Calibration of Dynamic Mechanical
Analyzers

ISO 6721-1 (2001)
Plastics – Determination of Dynamic Mechanical Properties – Part 1: General Principles

ISO 6721-2 (1995)
Plastics – Determination of Dynamic Mechanical Properties – Part 2: Torsion-Pendulum
Method / Note: Technical Corrigendum 1

ISO 6721-3 (1995)
Plastics – Determination of Dynamic Mechanical Properties – Part 3: Flexural Vibration –
Resonance-Curve Method / Note: Technical Corrigendum 1

ISO 6721-4 (1994)
Plastics – Determination of Dynamic Mechanical Properties – Part 4: Tensile Vibration –
Non-Resonance Method

ISO 6721-5 (2003) DAM 1
Plastics – Determination of Dynamic Mechanical Properties – Part 5: Flexural Vibration;
Non-Resonance Method; Amendment 1 /
Note: Intended as an Amendment to ISO 6721-5 (1996)

ISO 6721-6 (2003) DAM 1
Plastics – Determination of Dynamic Mechanical Properties – Part 6: Shear Vibration;
Non-Resonance Method; Amendment 1 /
Note: Intended as an Amendment to ISO 6721-6 (1996)

ISO 6721-7 (2003) DAM 1
Plastics – Determination of Dynamic Mechanical Properties – Part 7: Torsional Vibration;
Non-Resonance Method; Amendment 1 /
Note: Intended as an Amendment to ISO 6721-7 (1996)

ISO 6721-8 (1997)
Plastics – Determination of Dynamic Mechanical Properties – Part 8: Longitudinal and
Shear Vibration – Wave-Propagation Method

ISO 6721-9 (1997)
Plastics – Determination of Dynamic Mechanical Properties – Part 9: Tensile Vibration;
Sonic-Pulse Propagation Method

ISO 6721-10 (1999)
Plastics – Determination of Dynamic Mechanical Properties – Part 10: Complex Shear
Viscosity Using a Parallel-Plate Oscillatory Rheometer

prEN 6032 (1996)

Aerospance series – Fibre Reinforcrd Plastics – Test Method; Determination of the Glass
Transition Temperatures

DIN 29 971 (1991)
Aerospace – Unidirectional Carbon Fibre-Epoxy Sheet and Tape Prepreg – Technical
Specification

DIN 53 545 (1990)
Determination of Low Temperature Behaviour of Elastomers – Principles and Test
Methods

DIN 65 583 (1999)
Aerospace – Fibre Reinforced Materials – Determination of Glass Transition of Fibre
Composites Under Dynamic Load

Other Standards

ISO 527 (1994)
Plastics – Determination of Tensile Properties – Part 1-5

ISO 179 (1993)
Plastics – Determination of Charpy Impact Strength

ISO 604 (1993)
Plastics – Determination of Compressive Properties

DIN 50 035 (1989)
Terms and Definitions used on Aging of Materials

ISO 1172 (1996)
Textile-Glass-Reinforced Plastics – Prepregs, Moulding Compounds and Laminates –
Determination of the Textile-Glass and Mineral-Filler Content – Calcination Methods

Abbreviations Used

DSC	**Differential Scanning Calorimetry**	
DSC		Differential scanning calorimetry
		Heat flux DSC
		Power compensation DSC
TMDSC		Temperature-modulated DSC
c_{DSC}	[%]	Degree of curing
c_p	[J/(g °C)], [J/(g K)]	Specific heat capacity
H	[J], [J/g]	Enthalpy
ΔH	[J], [J/g]	Enthalpy change
$\Delta H_{m,f,S}$	[J], [J/g]	Melting enthalpy, heat of fusion
$\Delta H_{c,K}$	[J], [J/g]	Crystallization enthalpy, heat of crystallization
ΔH_r	[J], [J/g]	Reaction enthalpy, heat of reaction
$\Delta H_{m,f,S}^{0}$	[J], [J/g]	Melting enthalpy (heat of fusion) of a 100% crystalline material
w_c, α	[%]	Degree of crystallization, crystallinity
m	[g]	Mass
ρ	[g/cm³]	Density
P	[W]	Power
ΔP	[W]	Power difference
$\dot{Q}$	[W], [W/g]	Heat flux
$\Delta \dot{Q}$	[W], [W/g]	Heat flux difference
t	[s]	Time
T	[°C], [K]	Temperature
ΔT	[°C], [K]	Temperature difference
v_h	[°C/min], [K/min]	Heating rate (β_h)
v_k	[°C/min], [K/min]	Cooling rate (β_k)

Glass Transition Temperatures

T_g	[°C], [K]	Glass transition temperature
T_{mg}	[°C], [K]	Midpoint temperature
T_{ig}	[°C], [K]	Starting temperature
T_{eig}	[°C], [K]	Extrapolated starting temperature
T_{fg}	[°C], [K]	End temperature
T_{efg}	[°C], [K]	Extrapolated end temperature

Melting Temperatures

T_m	[°C], [K]	Melting peak temperature
T_m^0	[°C], [K]	Equilibrium melting temperature
T_{im}	[°C], [K]	Starting temperature
T_{eim}	[°C], [K]	Extrapolated starting temperature
T_{pm}	[°C], [K]	Peak temperature
T_{efm}	[°C], [K]	Extrapolated end temperature
T_{fm}	[°C], [K]	End temperature

(Crystallization and reaction temperatures are similar, but use subscripts c and r)

Temperature-Modulated DSC

$A(t)$	[°C/min], [K/min]	Heating rate amplitude (time-dependent)
A_{mod}	[°C/min], [K/min]	Modulation amplitude
$A_{mod.\Delta H}$	[W/g]	Amplitude of the modulated heat flux signal
$A_{mod.v}$	[°C/min], [K/min]	Amplitude of the modulated heating rate
P	[s]	Duration of period

OIT   Oxidative Induction Time/Temperature

Stat. OIT		Determination of the OIT time at constant temperature
Dyn. OIT		Determination of the OIT temperature on programmed temperature increase
t_U	[min]	Switching time from nitrogen to oxygen

t_{eio}	[min]	Extrapolated starting time of oxidation
t_x^{st}	[min]	Time after x [W/g] of exothermic shift from the baseline
T_{eio}	[min]	Extrapolated starting time of oxidation
T_x^{dy}	[min]	Temperature after x [W/g] of exothermic shift from the baseline

TG	**Thermogravimetry**	
DTG curve		Differential TG curve dm/dt
$A_{(1,\,2\ldots)}$		Starting point
$T_{A(1,\,2\ldots)}$	[°C], [K]	Temperature at starting point
$t_{A(1,\,2\ldots)}$	[min]	Time at starting point
$C_{(1,\,2\ldots)}$		Midpoint
$T_{C(1,\,2\ldots)}$	[°C], [K]	Temperature at mid-point
$t_{C(1,\,2\ldots)}$	[min]	Time at midpoint
$B_{(1,\,2\ldots)}$		End point
$T_{(B1,\,2\ldots)}$	[°C], [K]	Temperature at end point
$t_{(B1,\,2\ldots)}$	[min]	Time at end point
Subscrupt$_{1,\,2\ldots}$		1st and 2nd steps
m_s	[mg], [%]	Specimen mass
m_i	[mg], [%]	Midpoint between two decompositions
m_f	[mg], [%]	Mass after end temperature attained
m_{B1}	[mg], [%]	Mass at end point of the first loss of mass
m_{A2}	[mg], [%]	Mass at starting point of second loss of mass
m_{max}	[mg], [%]	Maximum mass occurring
Δm	[mg], [%]	Change of mass
$M_{L(1,\,2\ldots)}$	[mg], [%]	Loss of mass
M_G	[mg], [%]	Gain in mass
T_{p1}	[°C], [K]	1st peak maximum on the DTG curve
T_{p2}	[°C], [K]	2nd peak maximum on the DTG curve

TMA   **Thermomechanical Analysis**

l	[mm]	Length
l_0	[mm]	Starting length/reference length
Δl	[µm], [mm]	Change of length
$\Delta l/l_0$	[µm/m]	Change of length expressed in terms of the starting length
Δl_{th}	[µm], [mm]	Temperature-dependent change of length
T_0	[°C], [K]	Reference temperature
$\alpha(T)$	$[K^{-1}]$, $[10^{-6}K^{-1}]$, [µm/(m K)]	Differential coefficient of linear thermal expansion
	$[10^{-6}°C^{-1}]$, [µm/(m °C)]	
$\overline{\alpha}\,(\Delta T)$	$[K^{-1}]$, $[10^{-6}K^{-1}]$, [µm/(m K)]	Mean coefficient of linear thermal expansion
	$[10^{-6}°C^{-1}]$, [µm/(m °C)]	between two temperatures
$T_{g\alpha}$	[°C], [K]	Glass transition temperature from the α–curve
T_P	[°C, K]	Penetration Temperature
F		Calibration factor from $\alpha_{Experiment}$ and $\alpha_{Literature}$

DMA   **Dynamic Mechanical Analysis**

E	[MPa], [N/mm²]	Elasticity modulus
G	[MPa], [N/mm²]	Shear modulus
K	[MPa], [N/mm²]	Bulk Compression modulus
L	[MPa], [N/mm²]	Uniaxial-strain modulus
E* / G*	[MPa], [N/mm²]	Complex E-/G-modulus
E´ / G´	[MPa], [N/mm²]	Storage modulus
E´´ / G´´	[MPa], [N/mm²]	Loss modulus
δ	[rad], [°]	Phase angle
tan δ		Loss factor

f	[Hz], [s^{-1}]	Frequency
ω	[Hz], [s^{-1}]	Angular frequency
σ, σ_A	[MPa], [N/mm^2]	Stress, stress amplitude
ε, ε_A	[%]	Deformation, deformation amplitude
F	[N]	Force
M	[Nm]	Torque
τ	[MPa], [N/mm^2]	Shear stress
γ	[-]	Shear
k		Geometry factor
T_g	[°C], [K]	Glass transition temperature
T_{eig}	[°C], [K]	Extrapolated starting temperature
T_{mg}	[°C], [K]	Midpoint temperature
T_{fig}	[°C], [K]	Extrapolated end temperature
$T_{g2\%}$, T_{GA}	[°C], [K]	Start of glass transition temperature by the 2 % method
T_{g0}, T_W	[°C], [K]	Start of glass transition temperature by the tangents method
$T_g(E''_{max})$		
$T_g(G''_{max})$	[°C], [K]	Peak maximum temperature of the loss modulus
$T_g(\tan \delta_{max})$	[°C], [K]	Peak maximum temperature of the loss factor

µTA   Micro-Thermal Analysis

µTA^{TM}		Micro-Thermal Analysis
V	[V]	Voltage (Z Piezo element)

Abbreviations for Plastics

ABS	acrylonitrile-butadiene-styrene copolymer
AN	acrylonitrile
ASA	acrylonitrile-styrene-acrylate copolymer
B	butadiene
BR	butadiene rubber
CA	cellulose acetate
CAB	cellulose acetate butyrate
COC	cycloolefin copolymer
EP	epoxy resin
EPDM	ethylene-propylene-diene rubber
EPS	expanded polystyrene
EVAC	ethylene-vinyl acetate plastic
LCP	liquid crystalline polymer
MF	melamine-formaldehyde
mPE	metallocene catalyzed polyethylene
NR	natural rubber
PA	polyamide
PA 46	polyamide 46
PA 6	polyamide 6
PA 610	polyamide 610
PA 66	polyamide 66
PA 6-GF	glass-fiber reinforced polyamide 6
PA 66/ 6	polyamide 66/polyamide 6 copolymer
PA 6-3-T	amorphous polyamide
PAI	polyamide-imide
PAN	polyacrylonitrile
PAR	polyacrylate

PB	polybutene
PBI	polybenzimidazole
PBT	polybutylene terephthalate
PBT/PC	polybutylene terephthalate/polycarbonate blend
PBT-GF	glass-fiber-reinforced polybutylene terephthalate
PC	polycarbonate
PC+ABS	polycarbonate blend/acrylonitrile-butadiene-styrene copolymer
PE	polyethylene
PE-LD	polyethylene, low density
PE-LLD	polyethylene, linear low density
PE-HD	polyethylene, high density
PE-UHMW	polyethylene, ultrahigh molecular weight
PEEK	polyetherether ketone
PEI	polyetherimide
PEK	polyetherketone
PES	polyethersulfone
PET	polyethylene terephthalate
PF	phenolic resin
PI	polyimide
PIB	polyisobutylene
PMMA	polymethylmethacrylate
POM	polyoxymethylene
PP	polypropylene
PPE	polyphenyleneether modified
PP-GF	glass-fiber-reinforced polypropylene
PPO/PS	polyphenylene oxide/polystyrene blend
PPS	polyphenylene sulfide
PPSU	polyphenylene sulfone
PS	polystyrene

PS-HI (HIPS)	polystyrene high-impact
PS-S	polystyrene syndiotactic
PS-I	styrene-butadiene copolymer
PSU	polysulfone
PTFE	polytetrafluoroethylene
PUR	polyurethane
PVC	polyvinyl chloride
E-PVC	emulsion polyvinyl chloride
S-PVC	suspension polyvinyl chloride
PVAC	polyvinyl acetate
PVDF	polyvinylidene fluoride
S	styrene
SAN	styrene-acrylonitrile copolymer
SB	styrene-butadiene copolymer
SEBS	styrene-ethylene/butylene-styrene block copolymer
SI	silicone
TPE	thermoplastic elastomer
TPO	olefinic thermoplastic elastomer
TPU	thermoplastic polyurethane elastomer
UF	urea-formaldehyde polymer
UP	unsaturated polyester
VE	vinyl ester resin

Abbreviations for Plasticizers

DIDA	diisodecyl adipate
DIDP	diisodecyl phthalate
DOP	dioctyl phthalate
DOS	dioctyl sebacate
TCP	tricresyl phosphate

Chemical Structures of Low Molecular Weight Materials

Ar	argon
$CaCO_3$	calcium carbonate
CCl_4	carbon tetrachloride
CH_4	methane
CO_2	carbon dioxide
H_2O	water
H_2SO_4	sulfuric acid
HCl	hydrochloric acid
HCN	hydrocyanic acid
HNO_3	nitric acid
$KMnO_4$	potassium permanganate
N_2	nitrogen

1   Differential Scanning Calorimetry (DSC)

1.1   Principles of DSC

1.1.1   Introduction

Calorimetry is a technique for determining the quantity of heat that is either absorbed or released by a substance undergoing a physical or a chemical change. Such a change alters the internal energy of the substance. At constant pressure, the internal energy is known as enthalpy, H.

For practical applications, we are interested mostly in the change of enthalpy ΔH between two states:

$$\Delta H = \int c_p \cdot dT$$

Processes that increase enthalpy such as melting, evaporation or glass transition are said to be endothermic while those that lower it crystallization, progressive curing, decomposition are called exothermic, Fig. 1.1.

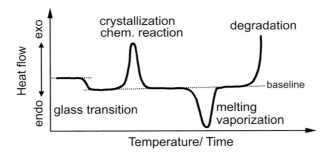

Fig. 1.1   Schematic diagram of a DSC curve showing possible transitions

The change in enthalpy is measured with the aid of a calorimeter by recording the displacement of the heat flux $\dot{Q}$ from a baseline. The baseline is a linear section of the curve that represents conditions in which no reaction or transition is occurring. It is assumed that the heat of reaction or transition or both is equal to zero at the baseline.

The specific heat capacity c_p is the quantity of energy needed to raise the temperature of 1 g of material by 1 °C at constant pressure. Because elaborate equipment is needed for measuring c_p, in DSC we determine the heat flux instead, that is the quantity of heat transferred per unit time and mass m. $\dot{Q}$ is directly proportional to the specific heat capacity. The proportionality factor is the heating rate v.

$$\frac{\dot{Q}}{m} = v \cdot c_P$$

This formula clearly shows the relationship between the most important influential factors, namely heating rate and mass.

Heat flow is measured as a function of temperature and/or time.

1.1.2 Measuring Principle

ISO 11357-1 [1] and DIN 51005 [2] identify two DSC methods for measuring the difference between thermal transitions in a specimen and those in a reference material:

      – Heat-flux DSC
      – Power-compensation DSC

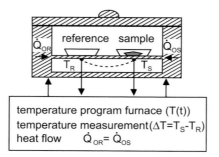

Fig. 1.2   Schematic diagram of a heat-flux DSC

      T_R = *Temperature of reference material*
      T_s = *Temperature of specimen*
      $\dot{Q}_{OR}$ = *Heat flow from furnace to reference pan*
      $\dot{Q}_{OS}$ = *Heat flow from furnace to specimen pan*

In **heat-flux DSC**, the test chamber consists of a furnace in which the specimen and reference material are heated or cooled together according to a controlled temperature program. The temperature of two points of measurement, which are located on a thermally conducting metal disc, is measured continuously. As long as the specimen and the reference material respond to the temperature program in the same way, heat flux into both the specimen $\dot{Q}_{OS}$ and the reference material $\dot{Q}_{OR}$ remains constant. Thus, the temperature difference between the two points of measurement is also constant.

For example, if ice were the specimen material, its temperature would remain at 0 °C while it was melting, despite the dynamic heating program. Hence the specimen temperature lags behind the reference material temperature until enough heat is added to bring about complete melting. Meanwhile, the reference material continues to heat up uniformly in accordance with the predetermined heating program. The difference between the two temperatures (ΔT) is the change in heat flux $\Delta \dot{Q}$.

The advantages of heat-flux DSC are primarily its relative robustness, ease of handling, and straightforward measurement, even in the case of offgassing specimens. The heating curves have a stable baseline and permit glass transitions to be measured very clearly.

In **power-compensation DSC**, the test chamber consists of two small, separate furnaces that are controlled independently by a defined primary heating program. If an exothermic or endothermic reaction in the specimen leads to a temperature difference ΔT between the two furnaces, power (energy) is applied to or removed from the specimen furnace to compensate for the energy change in the specimen. The system is ideally maintained in a "thermally null" state at all times. The difference in thermal power ΔP is the change in heat flux $\Delta \dot{Q}$ relative to the reference thermal power P_R.

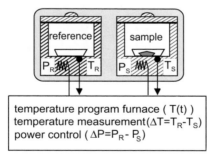

Fig. 1.3   Schematic diagram of a power-compensation DSC

T_R = Temperature of reference material
T_S = Temperature of specimen
P_R = Thermal power of reference furnace
P_S = Thermal power of specimen furnace

Power-compensation DSC can be used to measure very rapid reactions because the small furnaces have low time constants. As electrical compensation occurs very quickly, the temperature differences between specimen and reference material are very small.

Both processes are generally referred to as Differential Scanning Calorimetry (DSC). This is also the term used in [1]. As they yield virtually the same information for practical purposes, we will use the term DSC from this point on.

> **Heat-flux DSC and power-compensation DSC yield comparable information.**
> **Both are usually referred to as DSC.**

Tzero[TM] is an extension of classical h-f DSC which incorporates third thermocouple located on the heat flux plate so as to account for temperature gradients therein. A multi-term heat flow equation (which incorporates individual time constants and thermal resistances in the reference and sample sensors) is used to device the heat flow.

This procedure greatly improves the linearity of the baseline and thus the resolution of the effects being measured. Linear baselines with no drift allow dependable, reproducible evaluation. Consequently, unlike the situation for many other types of measurements, there is less need to subtract the baseline from the curve.

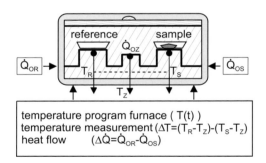

Fig. 1.4   Schematic diagram of Tzero[TM] sensor technology

T_R = Temperature of reference material
T_s = Temperature of specimen
T_z = Temperature of the Tzero sensor
$\dot{Q}_{OP}$ = Heat flow from furnace to specimen pan
$\dot{Q}_{OR}$ = Heat flow from furnace to reference material pan
$\dot{Q}_{OZ}$ = Heat flow from furnance to Tzero sensor

The technical capabilities of different DSC instruments are best assessed by comparing the following performance characteristics.

The smallest detectable signal is taken as a measure of the instrument's **sensitivity**. But before we can judge if an effect is significant, we need to know what the level of background noise is.   This is established by recording a baseline, either with or without the specimen. Sensitivity is usually expressed as effective noise in µW, Fig. 1.5.

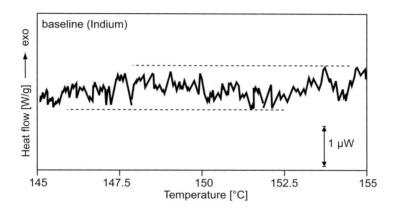

Fig. 1.5   Baseline noise (sensitivity) for a DSC experiment, indium

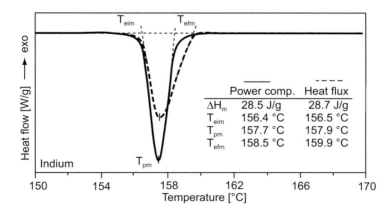

Fig. 1.6   Melting of a metal (indium), measured in a power-compensation DSC
                and a heat-flux DSC

ΔH_m = Heat of fusion

T_{eim} = Melting extrapolated onset temperature

T_{pm} = Melting peak temperature

T_{efm} = Melting extrapolated onset temperature

The **time constant** is a measure of the instrument's resolving power. It is a function of the thermal conductivity of the system, especially of the heating disc or the furnace and is expressed in seconds. Fig. 1.6 shows the melting of indium, measured in both heat-flux and power-compensation instruments.

The curves have different shapes. Curves measured by power-compensation DSC drop more steeply and have a narrower peak overall. Having passed through the melting temperature, the curve quickly returns to the baseline. The melting extrapolated end temperature T_{efm} (see Fig. 1.8) is lower than the T_{efm} of a curve recorded in a heat-flux calorimeter. The melting extrapolated onset temperatures T_{eim} and melting peak temperatures T_{pm} are almost identical on both instruments, as is the measured heat of fusion ΔH_m.

The time constant of power-compensation DSC is smaller because the furnaces are smaller and have a faster response, which makes them suitable for very rapid reactions. However, the lower sensitivity of instruments with a high time constant (> 10 s) can be an advantage if they are located in production areas where they are exposed to vibrations, and so forth.

1.1.3 Procedure and Influential Factors

The steps involved in conducting a DSC experiment are:

- Prepare the specimen.
- Weigh out the specimen into a pan.
- Seal the pan with the aid of a press.
- Place the specimen and reference pans into the test chamber.
- Set the flow of purge gas.
- Set the temperature program.

Factors influencing instruments and specimens are:

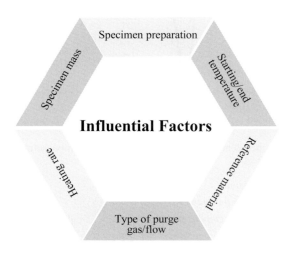

Influential factors and sources of experimental error are discussed in detail in Section 1.2.2 which includes curves from actual experiments.

1.1.4 Evaluation

DSC is suitable for measuring endothermic and exothermic effects that are due to an increase or decrease in enthalpy of the specimen. ISO 11357-1 [1] and ASTM D 3417-99 [3] recommend that endothermic effects be graphed on a negative y-axis, while DIN 53765 [4] recommends a positive y-axis. To avoid any confusion, we will use the expressions **endothermic** and **exothermic** instead of the algebraic sign convention.

1.1.4.1 Glass Transition

The glass transition of amorphous polymers or amorphous domains of semicrystalline thermoplastics marks the change from a glassy or energy-elastic state to a rubbery or entropy-elastic state. The mobility of the chain segments is greater above the glass transition temperature T_g, than it is below it (where it is said to be "frozen"). Because a new form of thermal mobility (segment movement) either occurs at the T_g, a step-like change occurs in the specific heat capacity c_p. The material undergoes a marked change in volume and enthalpy [5]. This is a relaxation transition and not a genuine phase transition. The temperature range over which it occurs is called the glass transition range or freezing range. By convention, the glass transition temperature T_g is that temperature at which half of the change in specific heat capacity has occurred [4].

Fig. 1.7 shows a typical glass transition for an endothermic process. A step-like change has occurred in the heat flux (or heat capacity). The various temperatures are defined below (see also [6, 7]).

Amorphous thermoplastics undergo major, usually step-like changes of properties in the glass transition range. In semicrystalline thermoplastics, these changes are less pronounced because of the presence of an as yet unmelted crystalline phase and they depend on the crystallinity, that is the proportion of crystalline phase.

The glass transition temperature T_g characterizes softening of the physical forces responsible for bonding in polymers. It is a function of the chemical structure and the extent of polymer branching and crosslinking. Above a certain molar mass, the T_g of thermoplastics is independent of the molar mass, but it is not independent of the degree of crosslinking in thermosets and elastomers. The shape and temperature position of the glass transition depend on the morphology of the polymer, while the morphology, in turn, is heavily dependent on the conditions employed in previous processing, for example cooling and freezing, and the thermal history of the polymer.

> **The temperature position of the glass transition depends on the processing-induced morphology (e.g., orientation, crystallization, crosslinking, internal stress).**

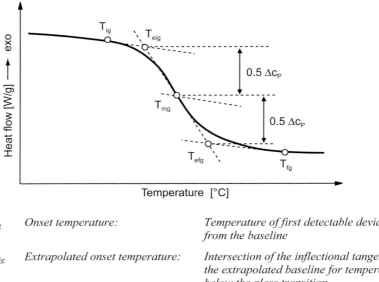

T_{ig}	Onset temperature:	Temperature of first detectable deviation from the baseline
T_{eig}	Extrapolated onset temperature:	Intersection of the inflectional tangent with the extrapolated baseline for temperatures below the glass transition
T_{mg}	Midpoint temperature, glass transition temperature:	Temperature at which half of the change in specific heat capacity ($0.5\Delta c_p$) has occurred; the temperature at which a line connecting the extrapolated baselines before and after the glass transition intersects with the DSC curve
T_{efg}	Extrapolated end temperature:	Intersection of the inflectional tangent with the baseline extrapolated from temperatures above the glass transition
T_{fg}	End temperature:	Temperature of the last detectable deviation from the baseline
Δc_p	Change in specific heat capacity:	Step height of endothermic glass transition

Fig. 1.7   Characteristic temperatures of a **glass transition**, based on ISO 11357-1 [1]

The point at which the inflectional tangent intersects with the heating curve is not always identical with the point at which the line connecting the baselines intersects with the heating curve. The reason is that it is not always obvious where to apply the tangents to the baseline. This difficulty is compounded by the fact that application of the tangents depends heavily on the choice of reference temperatures.

The authors in [8] and [9] cite repeatability values of 2.5 °C (two measurements in succession) and reproducibility values of 4.0 °C (measurements in different laboratories) for glass transition measurements.

A round-robin study performed on polystyrene using ten different DSC instruments yielded a glass transition temperature T_g of 107 °C with a range of +/- 2 °C. The same initial amount, heating rate, and evaluation method were used [10].

According to [4], the temperature values are to be quoted to the nearest 0.1 °C or 1 °C, the choice depending on the range of the thermal transition.

Our experience of the evaluation and repeatability of glass transition temperatures of polymeric materials is that temperatures should be quoted at best to the nearest 1 °C. While dedicated instrument software will return figures to the nearest 100th or 1000th of a degree, there is no metrological or processing basis for this and it serves only to create a false impression of the accuracy and specific property of a material. Such "accuracy" leads to confusion and causes major influential factors to be overlooked.

Temperatures characterizing the glass transition should be quoted to the nearest 1 °C.

National standards employ different terms to describe the various aspects of the glass transition step:

ISO 11357-1 [1]	ASTM D 3418 [8]	DIN 53 765 [4]	
T_{ig} Onset temperature	-	T_{gO} Onset temperature	[°C]
T_{eig} Extrapolated onset temperature	T_{eig} Extrapolated onset temperature	T_{gO}^E Extrapolated onset temperature	[°C]
T_{mg} Midpoint temperature	T_{mg} Midpoint temperature	T_g Glass transition temperature	[°C]
T_{efg} Extrapolated end temperature	T_{efg} Extrapolated end temperature	T_{gF}^F Extrapolated end temperature	[°C]
T_{fg} End temperature	-	T_{gE} End temperature	[°C]
Δc_p Specific heat capacity	-	Δc_p Specific heat capacity	[J/(g °C)]

Table 1.1   Terms used in various standards to describe characteristic values of the glass transition

1.1.4.2   Melting

Melting is a change from a solid, crystalline state into an amorphous liquid state. No loss of mass or chemical change occurs. It is accompanied by an endothermic enthalpy change. Figure 1.8 illustrates the characteristic temperatures, melting enthalpy, and the labels used in a DSC curve, as set out in ISO 11357-1 [1].

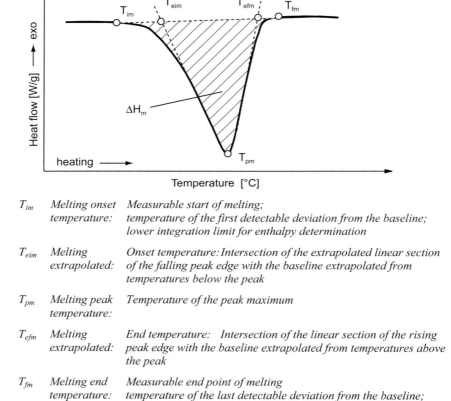

T_{im}	Melting onset temperature:	Measurable start of melting; temperature of the first detectable deviation from the baseline; lower integration limit for enthalpy determination
T_{eim}	Melting extrapolated:	Onset temperature: Intersection of the extrapolated linear section of the falling peak edge with the baseline extrapolated from temperatures below the peak
T_{pm}	Melting peak temperature:	Temperature of the peak maximum
T_{efm}	Melting extrapolated:	End temperature: Intersection of the linear section of the rising peak edge with the baseline extrapolated from temperatures above the peak
T_{fm}	Melting end temperature:	Measurable end point of melting temperature of the last detectable deviation from the baseline; upper integration limit for enthalpy determination; specimen in a liquid state
ΔH_m	Enthalpy change:	The quantity of heat absorbed (ΔH positive); The enthalpy change, according to [1] is denoted only by ΔH. To maintain consistent labeling, the subscript "m" has been introduced for the enthalpy change during melting.

The subscript "m" stands for melt, "i" for initial, "f" for final.

Fig. 1.8   Characteristic temperatures of a melting curve in accordance with ISO 11357-1[1]

Unlike metals, whose melting point represents the equilibrium temperature between solid and liquid, semicrystalline polymers melt over a relatively broad range. Like the glass transition range, the melting range is essentially governed by the structure of the polymers.

The actual melting process, the shape of the melting curve, and thus the characteristic values obtained from them depend heavily on the thermal and mechanical history of the specimen (see Section 1.2.3.3). Furthermore, the melting profile is influenced by the temperature-time regimen during heating (e.g., heating rate); low heating rates promote crystal reorganization or recrystallization or both in polymers, Fig. 1.35.

> **Melt profile is characterized by processing-induced morphology (e.g., orientation, crystallization) and the test conditions (e.g., heating rate)**

National standards employ different terms to describe the various characteristic temperatures of the melting process. An overview of the different terms and abbreviations is provided in Table 1.2.

ISO 11357-1 [1]	ASTM D 3417/3418 [3, 8]	DIN 53 765 [4]	
T_{im} Onset temperature	–	T_{SO} Onset temperature	[°C]
T_{eim} Extrapolated onset temperature	T_{eim} Melting extrapolated (fusion) onset temperature	T_{SO}^{E} Extrapolated onset temperature	[°C]
T_{pm} Peak temperature	T_{pm} Melting peak temperature	T_{SP} Peak temperature	[°C]
T_{efm} Extrapolated end temperature	T_{efm} Melting extrapolated end temperature	T_{SE}^{E} Extrapolated end temperature	[°C]
T_{fm} End temperature	–	T_{SE} End temperature	[°C]
ΔH_m Enthalpy change	ΔH_f Heat of fusion	ΔH_S Melt enthalpy	[J/g]

Table 1.2   Terms used in various standards to describe melting characteristics

In practice, a specific temperature, and not the complete melting curve, is used to describe the melting temperature T_m.

The fusion of metals (see Fig. 1.6) usually has a steep falling edge. Melting starts at the onset temperature T_{im} and continues until the entire specimen has melted at the melting peak temperature T_{pm}. Then the curve returns to the baseline (rising flank). This is where temperature compensation occurs between the specimen and reference materials. The melting temperature T_m reported is either the melting extrapolated onset temperature T_{eim} or the melting peak temperature T_{pm}.

Semicrystalline polymers consist of crystallites of different lamellar thicknesses and degree of perfection. The melting curve reflects this nonuniform structure, Fig. 1.9.

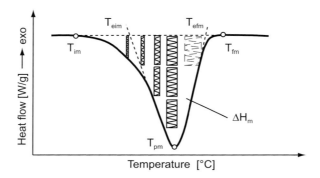

Fig. 1.9   Melting curve and lamellar thickness distribution for a semicrystalline
              thermoplastic

T_{im} = Onset temperature
T_{eim} = Extrapolated onset temperature
T_{pm} = Melting peak temperature
T_{efm} = Melting extrapolated onset temperature
T_{fm} = End temperature
ΔH_m = Enthalpy change

At the onset temperature T_{im} the thinner or less perfect crystallites, in as far as they can be detected by DSC, begin to melt. The melting peak temperature T_{pm} is the temperature at which most of the crystallites melt. It is difficult to say if the slope of the rising flank is due primarily to instrumental lag and thermal conductivity of the polymer specimen or to melting of thicker crystallites that are still present [11–13]. The latter is illustrated by the dashed line in Fig. 1.9. At the melting end temperature T_{fm}, all crystallites have definitely melted and the crystalline order has been destroyed. This temperature is also called the experimental melting temperature T_m [14].

The melting curve characterizes the lamellar thickness distribution; at T_{fm}, all crystallites have melted.

The temperatures listed so far are characteristic values that are dependent on various parameters. By contrast, the **equilibrium temperature** T_m^0 or thermodynamic equilibrium melting point is a material constant. T_m^0 is the melting point of a perfect, infinitely large crystal in which T_m and T_c are equal.

The equilibrium temperature T_m^0 is the melting point of a perfect, infinitely large crystal ($T_m = T_c$).

To determine this temperature by DSC, we start by measuring the difference in the melting and crystallizing temperatures in real crystallites. The specimen is melted and cooled from the melt to a certain temperature T_c called the isothermal crystallization temperature. After complete crystallization has occurred, the specimen is heated again and the melting temperature T_m of the resultant melting point curve is evaluated. T_{pm} or preferably T_{fm} can serve as T_m [16]. The measurements are repeated at different isothermal crystallization temperatures. Each melting temperature T_m is plotted against the crystallization temperature T_c, and a rectilinear relationship is obtained (dashed line in Fig. 1.10). This line is linearly extrapolated until it intersects with a line drawn through the origin representing $T_m = T_c$. The point where the two lines intersect is T_m^0. Fig. 1.10 shows how the equilibrium temperature is determined.

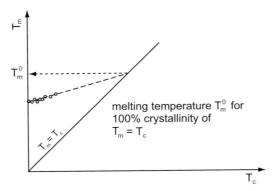

Fig. 1.10   Determination of equilibrium temperature, T_m^0, schematic

T_m = Melting temperature (either T_{pm} or T_{fm})
T_c = Isothermal crystallization temperature

Note: *Crystallization during the cooling phase must be avoided. This is accomplished by cooling as quickly as possible to the isothermal crystallization temperature T_c.*

The **enthalpy change ΔH** of a specimen, expressed in terms of the initial specimen mass [J/g], is calculated from the area bounded by the curve and the line connecting the onset temperature T_{im} to the end temperature T_{fm} (baseline for the integration). In the case of melting, this enthalpy change is called the heat of fusion ΔH_f, ΔH_m, or ΔH_S and is the energy needed to melt the existing crystalline fraction.

Heat of fusion: The energy needed to melt the crystalline fraction

The crystalline fraction may be expressed as a percentage of a value representing complete crystallization. This is the crystallinity, α. To avoid confusion with the coefficient of linear thermal expansion, α, we will use w_c to denote crystallinity.

The crystallinity of the specimen may be calculated from the experimental heat of fusion ΔH_m and the literature value for 100% crystalline material, ΔH_m^0.

$$w_c = \frac{\Delta H_m}{\Delta H_m^0} \cdot 100 \quad [\%]$$

Note: *There are different ways of determining the crystallinity of semicrystalline polymers, for example x-ray analysis, densitometry, calorimetry, and IR spectroscopy. The underlying assumption is that the crystalline and amorphous fractions have different properties. However, these methods do not detect orientation or transition states or, if so, to different extents. Furthermore, equilibrium states, crystalline modifications, and orientation may change during measurement. This explains the different values reorded for, for example, ΔH_m^0. The analytical method should always be recorded for this reason.*

Table 1.3 presents a summary of enthalpy values ΔH_m^0 and equilibrium melting temperatures T_m^0 as found in the literature. As there are no definitive values for ΔH_m^0 in the literature, the reference value should always be cited in addition to the crystallinity. Melting peak temperatures T_{pm} are listed for comparison. They were obtained on different specimens heated at a rate of 10 °C/min.

Material	ΔH_m^0 [J/g] [Wunderlich]	ΔH_m^0 [J/g] [van Krevelen]	ΔH_m^0 [J/g] [LKT]	T_m^0 [°C] [Wunderlich]	T_m^0 [°C] [LKT]	T_{pm} [°C] [LKT]
PE-LD	293	293	–	141	–	105–120
PE-HD	293	293	–	141	–	130–140
PP-H	207	207	205	188	191	160–165
POM-H	326	326	270	184	210 - 230	175–190
POM-C	–	–	220	–	180 - 190	140–170
PA 6	230	230	–	260	–	220
PA 66	255	300	–	301	–	260
PA 11	244	226	–	220	–	187
PA 610	284	208	–	233	–	222
PA 46	–	–	–	–	–	280–290
PET	140	145	–	280	–	240–260
PBT	140	–	–	248	–	220–230
PTFE	82	82	–	–	–	327

Table 1.3   Characteristic temperatures and enthalpy values for the crystalline fraction in semicrystalline thermoplastics [17–19]

T_{pm} from melting point curve (2nd heating scan), heating rate 10 °C/min, following cooling; cooling rate 10 °C/min

The **heat of fusion ΔH_m^0** of a 100% crystalline material may also be determined by DSC. In this case, we use the results from the determination of the equilibrium melting temperature, Fig. 1.10. The heat of fusion ΔH_m, obtained after isothermal crystallization, is plotted against the crystallization temperature T_c, Fig. 1.11. We then linearly extrapolate the values and read off the value that corresponds to the equilibrium melting temperature T_m^0 determined previously; this is the heat of fusion ΔH_m^0 of a theoretically 100% crystalline material.

Heat of fusion ΔH_m^0 is found at the equilibrium temperature T_m^0.

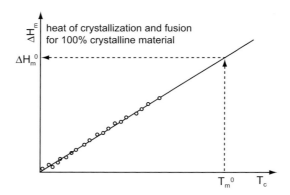

Fig. 1.11   Determination of the crystallization enthalpy ΔH_c^0 or heat of fusion ΔH_m^0 at the
           equilibrium temperature $T_m{}^0$, schematic

           T_c = Isothermal crystallization temperature
           ΔH_c^0 = Heat of crystallization for a 100% crystalline material
           ΔH_m^0 = Heat of fusion for a 100% crystalline material

In [8], 1.5 °C is cited for the repeatability (two measurements in succession) of the melting peak temperature T_{pm} and 2.0 °C, for the reproducibility (between different laboratories). A tolerance of +/- 5.5% is reported in [20] for the accuracy of the enthalpy determination.

It is recommended in [4] that temperatures be recorded to an accuracy of 0.1 or 1 °C, the choice depending on the width of the peak. Enthalpy values are recorded to the nearest 0.1 J/g or 1 J/g.

1.1.4.3   Crystallization

In DSC, the crystallization or cooling curve (exothermic) characterizes the change in enthalpy that occurs when, starting from high temperatures, a material in the liquid, amorphous state is transformed into a solid crystalline state, Fig. 1.12.

The characteristic temperatures are labeled as follows, according to draft standard ISO 11357-3 [21]; the standard ISO 11357-1 [1] describes these only for crystallization in the thermal (heating) curve.

Crystallization sets in at the onset temperature T_{ic} (Fig. 1.12) and cold crystallization (Fig. 1.13) becomes apparent. In practice, the crystallization extrapolated onset temperature T_{eic} is held to be the most important temperature on the cooling curve.

The crystallization peak temperature T_{pc} is the temperature at which the crystallization rate is a maximum. The crystallization extrapolated end temperature T_{efc} and the end temperature

T_{fc} characterize the detectable end of crystallization under the stated temperature – time conditions.

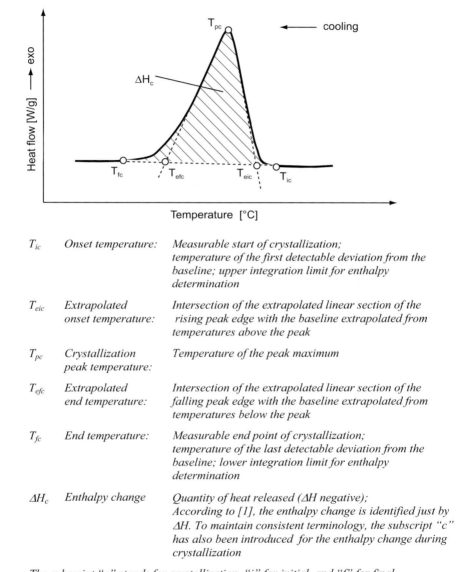

T_{ic}	Onset temperature:	Measurable start of crystallization; temperature of the first detectable deviation from the baseline; upper integration limit for enthalpy determination
T_{eic}	Extrapolated onset temperature:	Intersection of the extrapolated linear section of the rising peak edge with the baseline extrapolated from temperatures above the peak
T_{pc}	Crystallization peak temperature:	Temperature of the peak maximum
T_{efc}	Extrapolated end temperature:	Intersection of the extrapolated linear section of the falling peak edge with the baseline extrapolated from temperatures below the peak
T_{fc}	End temperature:	Measurable end point of crystallization; temperature of the last detectable deviation from the baseline; lower integration limit for enthalpy determination
ΔH_c	Enthalpy change	Quantity of heat released (ΔH negative); According to [1], the enthalpy change is identified just by ΔH. To maintain consistent terminology, the subscript "c" has also been introduced for the enthalpy change during crystallization

The subscript "c" stands for crystallization, "i" for initial, and "f" for final.

Fig. 1.12   Characteristic temperatures of a crystallization curve according to ISO 11357-3 [21] and adapted from ISO 11357-1[1]

ISO 11357-1 [1]*, [21]	ASTM D 3417/3418 [3, 8]	DIN 53 765 [4]	
T_{ic} Onset temperature	–	T_{KO} Onset temperature	[°C]
T_{eic} Extrapolated onset temperature	T_{eic} Crystallization extrapolated onset temperature	T_{KO}^E Extrapolated onset temperature	[°C]
T_{pc} Peak temperature	T_{pc} Crystallization peak temperature	T_{KP} Peak temperature	[°C]
T_{efc} Extrapolated end temperature	T_{efc} Crystallization extrapolated end temperature	T_{KE}^E Extrapolated end temperature	[°C]
T_{fc} End temperature	–	T_{KE} End temperature	[°C]
ΔH_c Enthalpy change	ΔH_c Heat of crystallization	ΔH_K Crystallization enthalpy	[J/g]

These terms were formulated from the standard as it provides an exact description of the crystallization peak only on heating, that is, starting from low temperatures.

Table 1.4   Terminology used in different standards to describe characteristic temperatures during crystallization

Onset and end temperature T_{ic} and T_{fc} can be measured only if the DSC instrument is sensitive enough to detect the thermal transition.

The onset and end temperatures T_{ic} and T_{fc} also depend on the sensitivity of the DSC instrument.

The cooling rate determines the position of the crystallization curve on the temperature scale and the characteristic values calculated therefrom. As the cooling rate increases, the crystallization curve is displaced toward lower temperatures.

The terms use to describe the various points on the crystallization curve differ somewhat in the various standards employed.

Crystallization from the melt can occur only after the temperature has fallen beneath the theoretical melting temperature T_m^0 (supercooling). Crystallization nuclei must also be present. These either form by themselves or are induced by deliberate addition of certain substances or both (nucleation, nucleating agents).

The nucleation and growth rates determine the overall crystallization rate recorded during a DSC experiment. Nucleation and growth occur in parallel and are heavily dependent on the degree of supercooling ΔT (T_m^0 - T_c).

DSC yields the mean overall crystallization rate.

The higher the cooling rate, the more extensive the supercooling. This improves the thermodynamic conditions for crystallization but impairs the kinetic conditions (molecular mobility). As a consequence, the overall crystallization rate (as well as the nucleation and growth rates) attains a temperature-dependent maximum between T_m^0 and T_g.

The higher the level of supercooling, the lower is the crystallinity of the melt. In fact, if the melt is quenched to temperatures below T_g, it freezes in a vitreous state. Slight supercooling through low cooling rates results in greater crystallinity.

In isothermal cooling, the lower the value of T_c, the lower is the attainable crystallinity w_c. In non-isothermal cooling, the conditions for nucleation and crystal growth are much more complicated, as they are continually changing [22].

The addition of nucleating agents increases the density of nuclei. Given the same cooling rate, crystallization will begin at higher temperatures, that is, a lower degree of super-cooling.

Crystallization can be measured not only during cooling but also during heating. This is called **cold crystallization** (Fig. 1.13) and it occurs if the material is rapidly cooled from the melt without crystallization occurring to a low temperature, usually below its glass tran-sition temperature.

The characteristic temperatures are, according to [1], labeled by analogy with Fig. 1.12, but starting from low temperatures.

The onset temperature T_{ic} characterizes the start of crystallization
on heating and cooling.

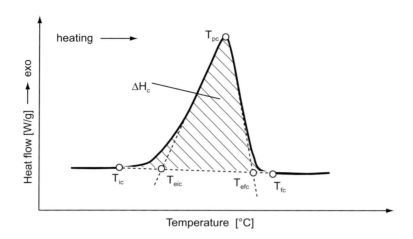

Fig. 1.13   Characteristic temperatures for a cold crystallization curve according to ISO 11357-1 [1]

The subscript "c" stands for crystallization, "i" for initial, and "f" for final.

1.1.4.4   Chemical Reaction — Dynamic Process

By analogy with the characteristic temperatures for melting and crystallizing, when a chemical reaction occurs, the labels are assigned the subscript "r", Fig. 1.14.

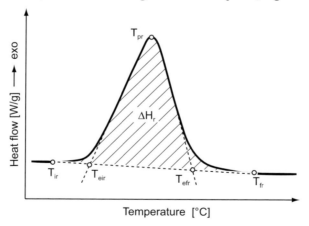

Fig. 1.14   Characteristic temperatures for a chemical reaction (here: curing of casting resins). DIN 53765 [4] describes the chemical reaction; temperatures are labeled in accordance with ISO 11357-1 [1], ISO 11357-5 [23]

The subscript "r" stands for reaction, "i" for initial temperatures, and "f" for final.

At T_{ir}, an identifiable chemical reaction is starting, and at T_{fr} it is finishing. The area ΔH_r characterizes the enthalpy change of an exothermic or endothermic reaction (reaction enthalpy).

1.1.4.5   Specific Heat Capacity

The specific heat capacity c_p at constant pressure is the quantity of heat that has to be applied to a material to raise its temperature by 1 °C. It is always required for calculations involving thermal processes.

To determine the specific heat capacity of an unknown material, we need to make three measurements. These are performed on:

- two empty pans,
- a pan filled with standard reference material and an empty pan,
- a pan filled with specimen material and an empty pan.

For the calibration, the first step is to establish the difference in heat fluxes (empty pans and standard reference material; the specimen usually is sapphire). The specific heat capacity of the unknown specimen is calculated from the difference in heat flux between the empty pan and the specimen material [24, 25]. The specific heat capacity c_p is calculated as a function of temperature from the following equation [26]:

$$c_p = \frac{\left[\dot{Q}\,(\text{specimen material}) - \dot{Q}\,(\text{empty pans})\right]}{v \cdot m}$$

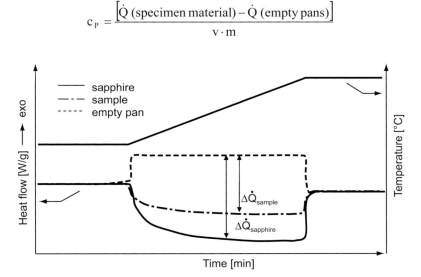

Fig. 1.15   Heat flow curves for determining the specific heat capacity in accordance with DIN 53 765 [4] and ISO 11357-5 [21]

The corresponding heat flux curves for determining the specific heat capacity by means of DSC are illustrated in Fig. 1.15.

Tables of c_p values are available for sapphire. Over the range 0–200 °C, the values change from 0.718 to 1.016 J/(g °C). For this reason it is advisable to make measurements over narrow temperature ranges or use a calibration factor that changes with temperature. A chemical reaction or physical change in the specimen must not occur in the temperature range being measured.

Direct measurement of the specific heat capacity after calibration is possible with TMDSC (see Section 1.1.6).

1.1.4.6   Test Report

ISO 11357-1 provides useful information on how to compile a complete test report that describes all experimental parameters and specimen details.

Where appropriate, the test report should contain the following information [1]:

- Reference to the standards employed,
- All information necessary for complete identification of the material analyzed,
- Type of DSC instrument used,
- Types of pans,
- Standard reference materials, their characteristics and the mass used in each case,
- Gas and gas-flow rate used,
- Details of sampling, preparation of the test specimen and the specimen-conditioning procedures used,
- Mass of the test specimen,
- Thermal history of the sample and the test specimen before the test,
- Temperature program parameters, including initial temperature, heating rate, final temperature, and cooling rate,
- Change of mass of the test specimen (if any),
- Test results,
- Date of the test,
- DSC curve.

1.1.5 Calibration

Measured temperatures and enthalpy changes need to be "true" values. A DSC instrument does not measure the actual temperature of the specimen directly and so there is a difference between the temperature of the specimen and the displayed value of the temperature measured directly beneath it. The latter has to be determined by calibration. However, because this difference is not linear over the entire temperature range, at least two standard reference materials appropriate to the temperature range must be used for the measurement. Calibration is affected by the [1]:

- Type of instrument used,
- Gas and its flow rate,
- Type of specimen pan used, its dimensions and its position in the specimen holder,
- Mass of the test specimen,
- Heating and cooling rates,
- Type of cooling system used.

It is advisable to calibrate the instrument in accordance with the manufacturer's instructions. Most recommended standard reference materials can be used for calibrating both the temperature and the enthalpy, Table 1.5.

Calibration parameters = Experimental parameters
At least two standard reference materials appropriate to the temperature range of the specimen are needed.

1.1.5.1   Temperature Calibration

The materials used for temperature calibration are chosen to suit the temperature range to be examined. Indium is suitable for calibrating the medium temperature range of many polymers. n-Heptane or water is suitable for low temperatures and lead or zinc may be used for high temperatures (see Table 1.5).

Comparisons can be made with literature values by extrapolating the results of dynamic measurements conducted at different heating rates to a heating rate of zero; this is a standard software feature on most instruments, Fig. 1.16 [27].

For gradual temperature gradients, for example in the case of highly pure metals, calibration is performed using the experimental melting extrapolated onset temperature T_{eim}; the peak maximum T_{pm} is used in the case of steeper temperature gradients or broader peaks [14, 28].

At least two standard reference materials should be measured over a temperature range suitable for the specimen. They should not be heated to more than 10 °C beyond their transition temperature in order that reactions between the specimen and standard pan materials may be avoided. Temperature calibration is always performed in the heating mode, to avoid complications due to supercooling.

For temperature calibration on cooling see [29].

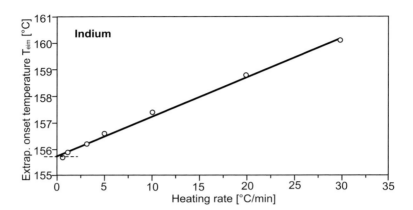

Fig. 1.16   Experimentally determined melting extrapolated onset temperatures T_{eim} of indium
               as a function of heating rate

Specimen mass 6.418 mg, purge gas: nitrogen

Reference material	Transition or melting point (equilibrium temperature) [°C]	Heat of fusion [J/g]
Cyclohexane (transition)	-83*	
Mercury (melting)	-38.9	11.47
1,2-Dichloroethane (melting)	-32*	
Cyclohexane (melting)	7*	
Phenylether (melting)	30*	
o-Terphenyl (melting)	58*	
Biphenyl (melting)	69.2	120.2
Potassium nitrate (transition)	127.7	
Indium (melting)	157.0	28.42
Tin (melting)	231.9	60.22
Lead (melting)	327.5	23.16
Zinc (melting)	419.6	107.38

Peak temperature

Table 1.5   Transition or melting temperatures and heats of fusion for various reference materials,
               as set out in ISO 11357-1 [1]

Reference material	Onset temperature T_{eim} [°C]	Midpoint temperature T_{mg} [°C]
Polystyrene	104.5	107.5

Table 1.6   Reference material for glass transition temperature, as set out in ISO 11357-1 [1]

1.1.5.2   Energy Calibration (Enthalpy Calibration)

Energy calibration is performed by means of substances of known heat of transition, for example indium, tin, or zinc, Table 1.5. The measured heat of transition is proportional to the area defined by the peak in the DSC curve and a virtual baseline (see Section 1.1.4.2).

The proportionality factor between the measured area and the heat of fusion of the standard reference material employed is also determined as a function of temperature. The resultant compensation curve returns a temperature-dependent calibration factor. At least two substances having different heats of transition should each be measured three times and the mean value taken [30, 31].

1.1.5.3   Heat-Flux Calibration by Means of Known Heat Capacity

In ideal conditions, the difference in true heat flux into the specimen and reference materials is the difference in heat capacities. The difference in true heat flux is determined from two experimental curves.

The first is obtained from measuring empty pans and the second, from a measuring a specimen (e.g., sapphire) and reference material. The difference between the two experimental curves is multiplied by the calibration factor (e.g., c_p of sapphire) to yield the true heat flux difference. Alternately, the measured difference in heat flux may be used to determine the difference in heat capacity [30].

At least two successive measurements are needed for calibration of heat flux and heat capacity.

Sapphire disks are suitable for measuring the specific heat capacity over a wide temperature range. When the temperature range is very wide, it must be divided into small steps for measurement and calibration. The reason is that the c_p values of sapphire vary with temperature [32].

In the following table c_p values of sapphire are shown as a function of temperature.

Temperature [°C]	c_p [J/(g °C)]	Temperature [°C]	c_p [J/(g °C)]
-103.15	0.3913	76.85	0.8713
- 83.15	0.4659	96.85	0.9020
- 63.15	0.5356	116.85	0.9296
- 43.15	0.5996	136.85	0.9545
- 23.15	0.6579	156.85	0.9770
- 3.15	0.7103	176.85	0.9975
0	0.7180	196.85	1.0161
16.85	0.7572	216.85	1.0330
36.85	0.7994	236.85	1.0484
56.85	0.8373	256.85	1.0627

Table 1.7   Specific heat capacity of sapphire (aluminum oxide), as set out in [33]

1.1.6 Temperature-Modulated DSC (TMDSC)

Temperature-modulated DSC is a special variant of differential scanning calorimetry. It can be used to distinguish thermally reversible transitions (glass transition, melting) from irreversible ones (crosslinking, decomposition, evaporation, cold crystallization, etc.). It permits some processes that either overlap or occur in rapid succession to be separated and makes it possible to glean information from poorly defined glass transitions, for example, in the case of semicrystalline thermoplastics.

The overall heat flux signal, corresponding to the heat flux signal of a conventional DSC experiment (cumulative curve), is split into reversing and non-reversing heat flux signals. The reversing signal is frequently also called the sensitive or repeatable signal. It can be reproduced by repeated heating, and is a function of the heat capacity and the heating rate. The non-reversing signal, said to be latent or non-repeatable, cannot be reproduced afterwards. Typical experimental curves produced by temperature modulated DSC are shown in Fig. 1.17.

> **Resolution of superimposed effects (reversing and non-reversing processes)**
> **Measurement of poorly defined glass transitions**
> **More accurate measurement of specific heat capacity**

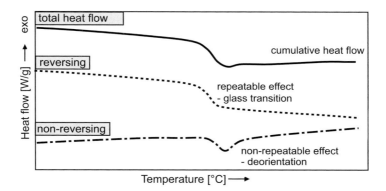

Fig. 1.17   Plot of a TMDSC experiment; overall heat flow, reversing and non-reversing heat flow

In the overall heat-flux curve, glass transition and deorientation are superimposed. This makes it difficult to evaluate T_g. In contrast, reversible heat flux has a well defined glass transition step that is easy to evaluate. In the irreversible heat flux curve, only the de-orientation is visible.

TMDSC differs from conventional DSC in that the specimen is subjected to an oscillating rather than constant rate of change of temperature. Sinusoidal, triangular, or saw-toothed heating may be employed. The average temperature changes linearly.

TMDSC: Periodic heating of the specimen

Figure 1.18 shows the modulated temperature as a function of time and the resultant heating rate. The curve is characterized and greatly influenced by the **underlying linear heating rate**, which in this case is 2 °C/min, the **amplitude** (+/- 0.5 °C/min) and the **frequency** (period 60 s $\approx$ 0.017 Hz) **of the modulated temperature**. This corresponds to an instantaneous modulated heating rate that oscillates by +/- 3.2 °C/min about the linear heating rate of 2 °C/min (Fig. 1.18, top).

The heating rate oscillation thus undergoes a change of algebraic sign; there are not only heating cycles (maximum heating rate: 5.2 °C/min) but also cooling cycles (minimum heating rate: -1.2 °C/min). Cooling phases are not always desirable; when melting processes in particular are being observed, the parameters are chosen such that the specimen is not cooled.

The maximum heating rate amplitude that occurs is therefore very important. It is made up of the linear heating rate component and the oscillation component. The heating rate amplitude is calculated as follows:

$$A(t) = v + A_{mod} \cdot \frac{2\pi}{P} \cdot \cos\left(\frac{2\pi}{P} \cdot t\right)$$

$A(t)$	=	*Heating rate amplitude (time dependent)*
P	=	*Period*
v	=	*Linear heating rate*
t	=	*Time*
A_{mod}	=	*Modulation amplitude*

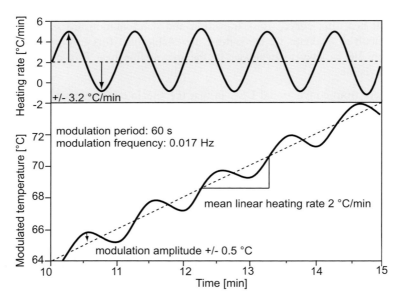

Fig. 1.18   Change in modulated temperature (below) and modulated heating rate (top) as a function of time

Small amplitudes are suitable for observing melting processes, and larger ones, for measuring glass transitions. The period should be chosen such that at least four cycles occur during a transition (e.g., glass transition). Although the experiments are more complicated than in standard DSC, they offer much better resolution and greater sensitivity. The slow, underlying, linear heating rate makes for high resolution and the steep periodic rise in heating rate increases the sensitivity.

Slow linear temperature rise: High temporal resolution
Steep periodic rise in heating rate: High sensitivity (strong signal)

The modulated heat flux signal is analyzed by a Fourier transformation to enable the overall heat flux and the reversible and irreversible signals to be derived.

The overall heat flux is a cumulative curve that corresponds to the heat flux of conventional DSC. To calculate the reversible heat flux, we need to find the specific heat capacity. We derive this from the data for the modulated heat flux and the modulated heating rate. The relationships used to calculate the specific heat capacity c_p are shown in Fig. 1.19 [33].

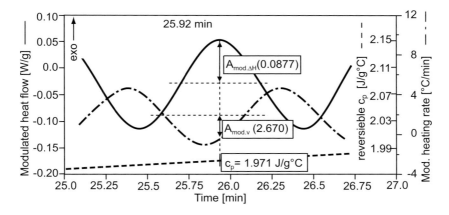

Fig. 1.19   Plot for determining the specific heat capacity c_p from the amplitude of the modulated heat flux and the amplitude of the modulated heating rate

Example: From the curves above, we can determine the **amplitude of the modulated heat flux signal** ($A_{mod.\Delta H}$ = 0.0877 W/g) and the **amplitude of the modulated heating rate** ($A_{mod.v}$ = 2.67 °C/min) at a specific measuring time of, say, 25.92 min. The ratio of these values, multiplied by a calibration factor (K), yields the heat capacity directly at this point in time: c_p = 1.971 J/(g °C). The calculation is shown below.

Multiplying the specific heat capacity by the mean heating rate yields the reversible heat flux. Subtracting the reversible heat flux from the overall heat flux yields the irreversible heat flux; note that this is not a calculated parameter. For further information about the complex mathematical relationships, please refer to [25, 34–36].

$$c_p = K \cdot \frac{A_{mod.\Delta H}}{A_{mod.v}} \quad \left[\frac{J}{g\,°C}\right] \quad \Rightarrow \quad c_p = 1 \cdot \frac{0.0877\,J \cdot 60\,s}{2.670\,°C \cdot s \cdot g} = 1.971 \quad \left[\frac{J}{g\,°C}\right]$$

$K \qquad = 1$

$A_{mod.\Delta H} = 0.0877\ [W/g] \qquad = 0.0877\ J/s \cdot g$

$A_{mod.v} \quad = 2.670\ [°C/min] = 2.670\ °C/60\ s$

At this time, there is still no uniform term to describe temperature-modulated DSC. The manufacturers sell their instruments under various names: TMDSC™, Temperature-modulated DSC; ADSC, Alternating DSC; DDSC, Dynamic DSC; and ODSC, Oscillating DSC.

1.1.7 Overview of Practical Applications

Table 1.8 indicates which DSC characteristics may be used to assess quality deficiencies, processing errors, and other parameters. These will be explained in Section 1.2.3 — Practical Examples.

Application	Characteristic	Example
Material identification, recipe components	T_{pm}, T_g	Distinguishing between polyamides (PA 6 and PA 66); incompatible mixtures and recyclates yield several melting peaks and T_g values
Modifications, additional components	T_g	Addition of plasticizers, Residual solvents, or monomers or the addition of plasticizers displaces the glass transition
Purity, impurities, compatibility, miscibility	T_{pm}, T_g	Residue of PE in PP molded part, silicone rubber in EP resin, T_g displacement
Process temperatures	T_{pm}, T_g, ΔH_m, c_p	Information about processing conditions, e.g. melt temperature, holding pressure, preheating temperature, welding temperature
Cooling conditions in the mold	T_{pc}, ΔH_c, ΔH_m	Information about freezing and mold temperatures, crystallization behavior, crystallinity

Application	Characteristic	Example
Evaluating thermal history	T_{pm}, T_g, ΔH_m	Information about overly-cold mold; recrystallization
Conditioning effects	T_{pm}	Identifying service temperatures
Crystallinity	ΔH_m	Information about component quality, mechanical properties
State/degree of curing	T_g, T_{pr}, ΔH_r	Displacement of glass transition, position and extent of residual reaction
Curing conditions	T_g, T_{eir}, ΔH_r	Start of curing, optimizing curing parameters through appropriate temperature and time
Storage stability, reactivity of individual components	ΔH_r	Reaction enthalpy required after protracted storage

Table 1.8   Some practical applications of DSC experiments on plastics, along with relevant characteristics.

1.2   Procedure

In this section, we will concentrate primarily on experimental parameters. When these are chosen carefully, the DSC results are useful and informative. Attention will be drawn to the possible errors associated with handling the specimen and the instrument, and with experimental parameters. The last part of this section deals with the vast investigative scope of thermal analysis.

Cross-references are provided to help you keep track of the underlying relationships.

1.2.1 In a Nutshell

Specimen preparation   The choice of **sampling site** and proper preparation of the very small specimen are crucial. Specimens from molded parts or pellets must be removed as carefully as possible with a scalpel. The sampling site is always chosen on the basis of the nature of the problem and must be documented.

Specimen mass   The **specimen mass** depends on the thermal transition to be measured. For melting and crystallization studies, it is 5–10 mg, although 1–5 mg is sufficient for highly sensitive instruments. A specimen mass of 10–20 mg is recommended for measurement of glass transitions. Measurements of specific heat capacity requires more material (20–40 mg). According to [1], deviations in specimen mass may not exceed +/- 0.01 mg.

> **Melting/crystallization processes: 1–10 mg**
> **Glass transition: 10–20 mg**

Pan   Pans come in different materials and shapes; disposable **aluminum pans** are the most commonly employed types. Pans and lids should be weighed to the nearest 0.01 mg, according to [1].

Purge gas   To avoid reactions with the environment, the chamber is purged with inert gas during the experiment. The gas may be nitrogen, helium, or argon. Oxygen would promote oxidation.

Temperature program   The temperature program is chosen to suit the material and specimen, and employs selected temperature limits at an appropriate heating rate.

Starting temperature   The **starting temperature** should be at least 50 °C below the temperature of the first expected transition; it must be maintained until the apparatus has reached equilibrium. At ambient temperature, an equilibration time of 5 min is adequate. At -100 °C, about 20 min are required.

The temperature and heat flux signal should be monitored constantly.

Heating rate

The choice of **heating rate** depends on the transition to be measured. For melting and crystallization studies, good results are obtained at 10 °C/min (or higher, for suppressing recrystallization). For glass transitions, 20 °C/min is adequate.

> **For melting/crystallization processes: 10 °C/min**
> **For glass transitions: 20 °C/min**

In **isothermal measurements**, the specimen is rapidly heated or cooled to a specific temperature and measured isothermally. The heating rate for achieving the isothermal holding temperature should be as high as possible but, to prevent "overshooting", it should not exceed 50 °C/min. When isothermal crystallization is being studied, the specimen must be cooled very rapidly to prevent crystallization from occurring during the cooling phase.

The test chamber may be heated to the desired temperature without specimen, for example, to track the curing of a resin system, and then the specimen is introduced. Introducing the cold specimen into the warm test chamber causes marked "undershooting" that hampers evaluation.

End temperature

For a dependable evaluation, the **end temperature** in the determination of glass transitions should be approx. 50 °C higher than that of the transition. When semicrystalline thermoplastics are being melted, heating must continue beyond the end temperature T_{fm} (Fig. 1.6) of the melting process so as to eliminate the influences of thermal and mechanical history. Limiting the end temperature to 30 °C above T_{fm} will rule out the possibility of specimen decomposition. If further temperature program steps are to follow, the end temperature should be maintained for about 2 min to allow thermal equilibrium to become established.

Cooling rate

Usually, the **cooling rate** should be the same as the heating rate, especially if the specimen is cooled to the starting

	temperature and then heated up again. The cooling rate is responsible for the "new" history.
1st heating scan	The **1st heating scan**, also called the initial run, reveals information about the current condition of the specimen, for example, the thermal and mechanical history (processing influences, crystallinity and curing, service temperatures). Subsequent, controlled cooling creates a "new", known specimen history.
2nd heating scan	Information yielded by the **2nd heating scan**, or 2nd run, is used for determining characteristic properties of the material.
3rd heating scan	In case of reactive resins for the validation of results obtained from the 2nd heating scan, especially with slowly curing resin systems where a high curing degree is expected.

> **1st heating scan:**
> **Current specimen condition (thermal and mechanical history + characteristic properties of the material)**
> **2nd heating scan:**
> **Only characteristic properties of the material (after controlled cooling)**

Evaluation/ Interpretation	When the DSC experiment has been successfully completed under using favorable experimental parameters, the transitions can be evaluated and interpreted properly. Interpretation presupposes a detailed knowledge of the material, processing conditions, and specimen preparation.

1.2.2 Influential Factors and Possible Errors During Measurement

1.2.2.1   Specimen Preparation

Specimen preparation consists of preparation, choice of sampling site, and pretreatment of the specimen.

Liquid or powdery specimens are transferred to the pan with the aid of a pipette, syringe, spatula, or toothpick. As they can be readily spread, good contact with the bottom of the pan is ensured. With liquid specimens, it is advisable just to lay the pan lid on top because the liquid might otherwise be squeezed out of the pan. Pellets, molded parts, or large finished parts require more elaborate sampling techniques.

Specimen preparation should not damage, deform, or heat the material. Soft materials are best prepared with a scalpel or punch, and hard materials, with a water-cooled saw or diagonal-nosed cutting pliers. Figure 1.20 shows the influence of specimen deformation on the heat of fusion of a polyethylene slab; an approx. 5 mm thick PE slab was clamped firmly in a vise and twisted forcibly with pliers. The specimen was removed by scalpel from both unstressed base material and twisted and crushed areas. Apart from the 15% reduction in heat of fusion, the reproducibility of the peak temperature of the crushed and twisted specimens was 5% poorer than that of the base material. Obviously the crystalline structure had been affected.

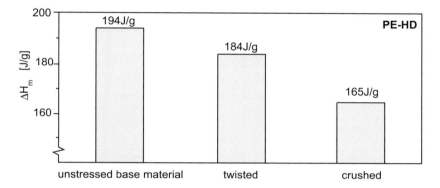

Fig. 1.20   Influence of mechanical preparation on the heat of fusion ΔH_m of a PE-HD lab

1st heating scan, specimen mass approx. 3 mg, heating rate 10 °C/min,

purge gas: nitrogen

To illustrate further the influence of preparation, specimens of an unfilled EP resin, cured for 2 weeks at ambient temperature, were removed both with a diagonal-nosed cutting pliers and by hammer, and then prepared, Fig. 1.21. The specimen removed with the pliers was sealed as a single piece in the pan and measured. Several transitions were revealed: the glass transition with subsequent endothermic effect and a residual exothermic reaction. The endothermic effect makes it difficult to apply the tangent to evaluate the T_g and to estimate the residual reaction. The second specimen, which had been hammered into several small pieces and then prepared, failed to exhibit an endothermic peak. Presumably, the different shape of its glass transition is due to de-aging arising from mechanical action.

Preparation work can greatly affect the results.
Specimen preparation should therefore be as gentle as possible.

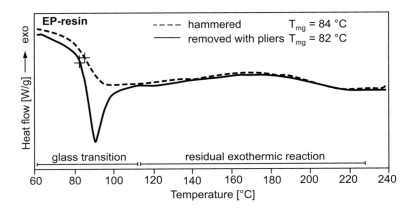

Fig. 1.21   Influence of preparation on the curve of an EP-resin specimen

T_{mg} = Glass transition temperature (midpoint temperature), 1st heating scan, specimen mass approx. 19 mg, heating rate 20 °C/min, purge gas: nitrogen

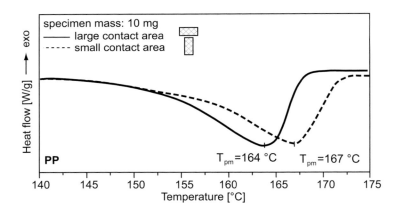

Fig. 1.22   Influence of specimen contact area on the melting point curve of PP

T_{pm} = Melting peak temperature, 1st heating scan, specimen mass approx. 10 mg, heating rate 10 °C/min, purge gas: nitrogen

DSC analysis relies upon good heat transfer between the specimen, pan, and test chamber. Heat transfer is enhanced by a specimen with a large **contact area** in the pan.

Figure 1.22 shows the heating curves for two identically sized PP specimens weighing approx. 10 mg, one placed horizontally and the other vertically in the pan. Heat transfer to the horizontal specimen is good. The vertical specimen has a smaller contact area and so it takes longer for all of it to melt. The melting peak temperatures T_{pm} for the 1st heating curve differ by 3 °C. When both specimens have completely melted, the material is uniformly dispersed in the pan and so there is no difference in the 2nd heating curves for the specimens.

Large contact area promotes heat transfer.

Next in importance after careful sampling is the choice of **sampling site** on the finished part. The results yielded by injection molded parts, in particular, vary considerably according as the site is close to/away from the gate or mold surfaces. Design influences (section thickness); different dwell times in the mold (flow paths, weld lines); and variation in holding pressure, mold, and melt temperatures cause extensive localization of crystallization behavior. Figure 1.23 shows the melting behavior of the surface of a PE beverage crate, both close to the sprue and away from it (at the handle).

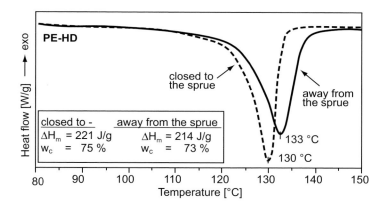

Fig. 1.23   Sampling close to and away from the sprue for a PE-HD beverage crate

T_{pm} = Melting peak temperature, ΔH_m = Heat of fusion
w_c = Crystallinity (expressed in terms of ΔH_m^0 = 293 J/g),
1st heating scan, specimen mass approx. 3 mg, heating rate 10 °C/min,
purge gas: nitrogen

A striking feature is the difference in peak positions. The specimen close to the sprue begins to melt at a lower temperature, with the melting peak temperature T_{pm} 3 °C lower than that of the specimen taken away from the sprue. The region close to the sprue was under pressure all the time the mold was being filled and this elevates the melting point. The effective supercooling ΔT is thus greater and leads to the formation of thinner lamellae that melt at lower temperatures (see Sections 1.2.3.1 and 1.2.3.5). In the region away from the gate, although the material cooled faster and thus crystallized at a lower temperature range, it underwent less supercooling. The heats of fusion are only slightly different.

Important information about the processing history, other than that yielded by specimens taken close to/away from the gate, is provided by measurements across the specimen's crosssection.

Figure 1.24 illustrates this for a molded POM part having a section thickness of approx. 3 mm. The material was injection molded and the specimen was removed at the surface and the middle, as viewed across the crosssection. The recorded melting point curves differ in their heats of fusion and resultant crystallinity. The specimen removed close to the surface has a heat of fusion and crystallinity roughly 10% lower because it cooled down rapidly at the cold mold wall. The width of the curves is also different. The overall effect is that of crystallites with broader lamellae forming in the center of the specimen.

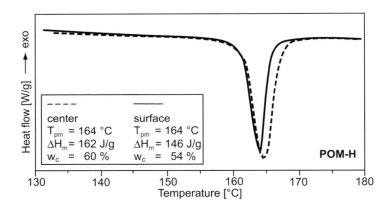

Fig. 1.24   Influence of sampling site (surface, center) on the melting point curve of POM-H

T_{pm} = Melting peak temperature, ΔH_m = Heat of fusion,
w_c = Crystallinity (expressed in terms of ΔH_m^0 = 326 J/g),
1st heating scan, specimen mass approx. 3 mg, heating rate 10 °C/min,
purge gas: nitrogen

Carefully choose and accurately document the position of the sampling site on the molding.

The following two typical examples illustrate the extent to which **specimen conditioning** affects the measured characteristic values and can lead to false interpretations.

Absorption of moisture by polyamide through attachment of water to the polar amide groups (-NHCO-) in the less densely packed amorphous domains leads to changes in electrical, mechanical and thermal properties (see sections on TMA, DMA, and TGA).

Plastics can absorb water (e.g., polyamides) and this affects their properties.

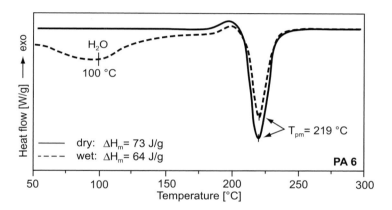

Fig. 1.25   Influence of moisture on PA 6

ΔH_m = Heat of fusion, T_{pm} = Melting peak temperature,
1st heating scan, specimen mass approx. 4 mg, heating rate 10 °C/min,
purge gas: nitrogen

Water absorption shows up on the heating curve in the form of a "hump", Fig. 1.25. As heating progresses, water begins to evaporate at around 30 °C, and this causes an endo-thermic change in the heating curve. Only when the temperature is roughly 100 °C does the baseline approach its original value. Because of the particularly strong bond between the water and the compact form of the specimen, such a hump is expected up to temperatures exceeding 150 °C, especially because the water cannot diffuse unhindered out of the specimen. The occluded water increases the original specimen mass and thereby falsifies the enthalpy calculation. To prevent the evaporating water from causing the pressure to build up, a small hole (0.5 mm max.) is punched in the lid of the pan. In the example under dis-cussion, two identically processed polyamide specimens were conditioned saturated (8%

H_2O) and dry (0.1% H_2O). As the water absorption can hardly affect the process-related crystallinity, the 12% lower heat of fusion ΔH_m of the wet specimen is due to the absorbed water.

Volatile substances such as water and solvents should be removed by drying or evaporation prior to measurement. To check if weight loss has occurred during the DSC experiment, it is best to weigh the pan plus specimen before and after measurement. Further consequences of water absorption (see Section 1.2.3.4) are depression of the glass transition and a change in the rigidity of the specimen with increase in temperature (see Sections 4.2.3.2 and 5.2.3.2).

A widespread application of DSC is to determine the **degree of curing of reactive resins** (see Section 1.2.3.8). First, the crosslinking enthalpy of the freshly prepared resin-curing agent mixture is measured so as to serve as the initial value. The mixture cures completely during the 1st heating scan in the DSC. The initial value is then compared with the experimental enthalpies that are obtained under different postcuring conditions.

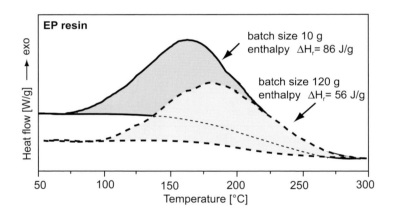

Fig. 1.26   Influence of batch size on the residual enthalpy ΔH_r of EP resin after one day at 23 °C

1st heating scan, specimen mass approx. 20 mg, heating rate 20 °C/min,
purge gas: nitrogen

Batch size of a resin-curing agent mixture affects the curing behavior.

Figure 1.26 shows the effect of the size of the batch containing the reactive agents. Different batches (10 g and 120 g), prepared with the same mixing ratio, were conditioned for one day in standardized conditions (23 °C/50% r.h.). The difference in enthalpies revealed by DSC experiments stemmed from the different extents of internal warming in the exothermic reaction. Internal warming is much more pronounced in the large batch than in the small one. It boosts curing over the course of one day and, in subsequent DSC experiments, resulted in a lower residual enthalpy as well as a delay in postcuring. This example also highlights the problem associated with scaling up laboratory results to real-life batches.

1.2.2.2 Specimen Mass

As a result of the physical relationships involved, the specimen mass has a considerable influence on the heat flux signal (see Section 1.1.1). Overloading effects and temperature gradients in the specimen should serve as a warning against thinking "a little extra won't do any harm". Instead, the specimen mass should be appropriate to the transition under measurement. Guide values are [4]:

Thermal transition	Specimen mass
Glass transition	10 to 20 mg
Melting/ crystallizing	5 to 10 mg*
Chemical reaction	10 to 20 mg
Specific heat capacity	20 to 40 mg

*In the author's experience, a specimen mass of 1 – 2 mg is adequate, provided the instrument is sensitive enough.

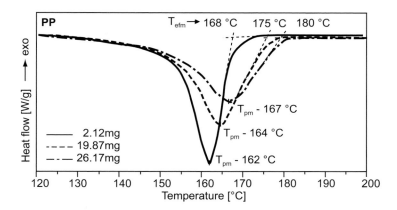

Fig. 1.27   Influence of specimen mass on the melting peak of PP

T_{pm} = Melting peak temperature,
T_{efm} = Extrapolated final temperature,
2nd heating scan, heating rate 10 °C/min, purge gas: nitrogen

With semicrystalline thermoplastics, too high a specimen mass can cause the melt to escape from the pan. This shows up as an unsteady baseline after the melting peak. Delayed heat conduction due to high specimen mass broadens the melting peak and displaces the melting peak temperature T_{pm} toward higher values, without changing the heat of fusion ΔH_m or the melting extrapolated onset temperature T_{eim}, Fig. 1.27.

The rising flank of the melting peak is influenced both by an instrumental effect, namely temperature compensation between the specimen and the reference material, and by a specimen effect, namely heat transfer in the specimen (see Section 1.2.3.1). High specimen masses and thus increased lag cause the rising flank to flatten out more and more, and the peak becomes broader.

Figure 1.28 is a plot of the width of the rising flank, that is the difference between melting extrapolated end temperature T_{efm} and melting peak temperature T_{pm}, as a function of specimen mass for the PP discussed above and for PBT. In these two specimens, there is an almost linear relationship between the temperature difference and the specimen mass.

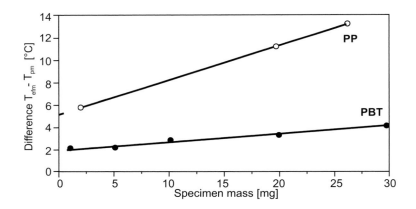

Fig. 1.28   Temperature difference between melting extrapolated end temperature

T_{efm} and melting peak temperature T_{pm} as a function of heating rate for PP and PBT

To be able to follow melting processes as precisely as possible and to minimize lag and heat conduction effects, it is best to keep the specimen mass at 1–2 mg, provided the instrument is sensitive enough.

When studying melting, keep specimen mass as low as possible.

Higher specimen masses are advisable for studying **glass transitions**. Figure 1.29 shows how the glass transition temperature T_{mg} varies with specimen mass in a polycarbonate and a completely cured epoxy resin. The glass transition temperature of the less thermally conductive epoxy resin (relative to PC) varies much more with increase in specimen mass than that of PC. Consequently, identical specimen masses must be used for comparative measurements.

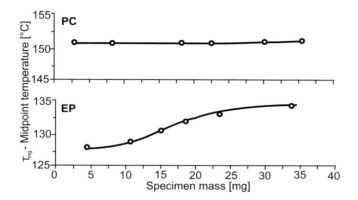

Fig. 1.29   Influence of specimen mass on T_{mg} (midpoint temperature) for PC and EP

2nd heating scan, heating rate 20 °C/min, purge gas: nitrogen

Comparative measurements must be performed on identical specimen masses.

The **fillers and reinforcing agents** added to the plastics affect the specimen mass and thus any quantitative information, for example via heat of fusion or reaction enthalpy, that is expressed in terms of the specimen mass. Precise knowledge of the proportion of reactive polymer in the pan is required in particular for thermosets, which are usually highly filled with fillers and reinforcing agents. If the filler distribution is both known and homogeneous, the specimen mass may be adjusted prior to the measurement.

The most reliable way of determining the exact specimen mass is to follow the DSC experiment with a thermogravimetric analysis, which decomposes the DSC specimen and permits the polymer fraction in the DSC specimen to be determined.

Comparisons of specimens of different weight are facilitated by plotting the heating curves in standardized form, that is heat flux in W/g against specimen mass.

For filled and reinforced polymers, the specimen mass must be corrected to reflect the composition.

1.2.2.3   Pan

For routine measurements (prescribed in [4]), the lid should be perforated to prevent pressure buildup. Although pre-perforated lids may be purchased, the perforation can be made with a needle. Where specimens are highly volatile (e.g., aqueous solutions), the size of the perforation is critical. In such cases, pre-perforated lids should be used.

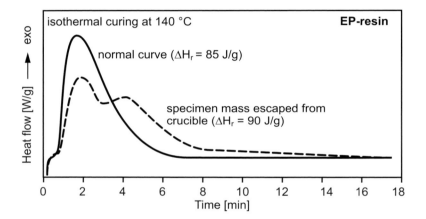

Fig. 1.30   Isothermal curing of EP-resin – curve for material escaped from the crucible and the normal curve

Specimen mass approx. 5 mg, isothermal 140 °C/min, purge gas: nitrogen

Where perforated lids are used, there is the risk that the material will escape and that the results will thereby misrepresented. This effect is shown in Fig. 1.31 for a glass-fiber-reinforced PA 66. Crucibles should therefore be closely inspected after each scan. Figure 1.30 shows the actual curve for an isothermal curing reaction and the curve for escaped specimen.

Standard situation: Aluminum pan with perforated lid

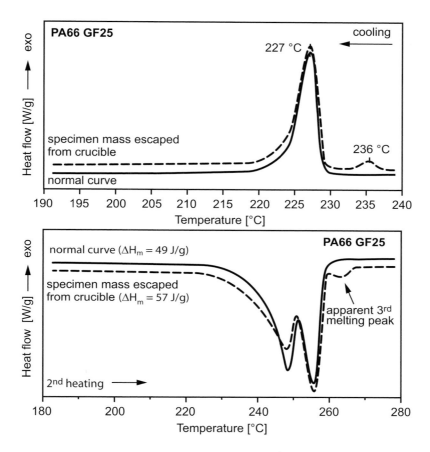

Fig. 1.31   Crystallization curve (left) for PA 66 GF25 and 2nd heating scan (top)
for PA 66 GF25, showing a curve for material escaped from the crucible
and the normal behavior

Specimen mass approx. 3 mg, heating rate 10 °C/min, purge gas: nitrogen

*Note: The lower the viscosity of the material at elevated temperatures, the higher the risk
of specimen escaping from the crucible*

Various types of pans are available, differing in material, volume, and specific use (e.g.
pressure pans). Disposable aluminum pans are generally used for analyzing plastics at
temperatures up to 600 °C because they generally do not react with the plastic under test.
There have been isolated reports of nonreproducible transitions that were caused by release

and drawing greases used in pan manufacture. This may be prevented by cleaning the pans with acetone or methanol. According to ISO 11357-1 [1], the pans and lids should be weighed to the nearest 0.01 mg and transferred with tweezers to the specimen holder in the test chamber.

Quartz, noble metal, or oxide ceramic pans are used for temperatures above 600 °C. Specimen volumes are 10–50 µl or, where large capacity pans are available, 75–100 µl. Standard practice is to crimp the lid against the pan. Such pans can withstand internal pressures of 2–4 bar.

Pressure pans are used for chemical reactions that either generate a product, for example, water released during the cross-linking of phenol resin, or involve a volatile reagent, for example, styrene in UP resin. They are sealed with a special tool and can resist internal pressures of roughly 150 bar. It should be borne in mind that pressure may build up in the pan and that the concentration of readily volatile reagents during a reaction does not change (see Section 1.2.3.8). Very low heating rates are necessary because the pans have to be thick (they are roughly 10 times as heavy as normal pans) and heat transfer will be sluggish.

Use a pressure crucible if readily volatile components are present.

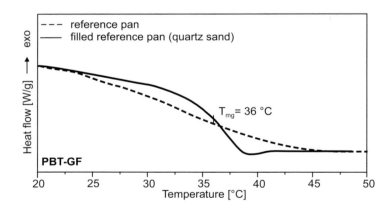

Fig. 1.32   Influence of filling the reference pan on the shape of the heating curve for the
                glass transition of PBT GF30

                2nd heating scan, specimen mass approx.10 mg, heating rate 20 °C/min,
                purge gas: nitrogen

An empty pan is usually placed in the **reference chamber** of the DSC instrument. Reference materials are used when the thermal transition is very small and the specimen mass cannot be increased further due to high filler content. Aluminum oxide and ground quartz glass may serve as reference materials for temperatures up to 600 °C. Ideally, the reference material is pure filler; any thermal transitions that occur in it have to be determined in a separate experiment.

Figure 1.32 shows that the glass transition of a fiberglass-reinforced PBT GF30 (poly-butylene terephthalate) measured against a filled reference pan is easier to identify and evaluate than an empty one. The reference pan was filled with an amount of quartz sand roughly corresponding to the fraction of fiberglass in the polymer.

1.2.2.4   Purge Gas

To prevent reactions (oxidation) between the specimen and its environment, DSC experiments must be performed in an atmosphere of highly pure (99.999%) inert gas, such as nitrogen, helium, or argon.

The flow of gas must be limited by a pressure reducer to roughly 1 bar and adjustable by means of a flowmeter. A flow rate of roughly 20 ml/min is recommended in [4]. Because chamber sizes vary from instrument to instrument, the manufacturer's instructions should be observed. Our studies have shown that peak temperature and heat of fusion are only slightly dependent on the gas pressure and thus on the flow rate; however, a certain minimum pressure is needed to purge the chamber completely.

The purge gas pressure does affect measurements of specific heat capacity c_p, which by definition varies with the pressure.

> **Purge gas pressure has little effect on T_{pm} and ΔH_m, but a greater effect on c_p.**

To prevent the purge gas from creating temperature gradients, it should be heated before entering the chamber. Nitrogen is generally chosen for temperatures ranging from ambient to roughly 600 °C. Its low cost is also a factor.

Helium is most often used for low-temperature experiments because its good heat conduction at low temperatures hastens the attainment of a constant temperature in the chamber. In our experience, a constant starting temperature of -100 °C can be reached in about 10 min with helium whereas about 30 min are needed if nitrogen is used.

It is crucial to use the same gas for both the calibration and the experiment, as otherwise enthalpy differences of up to 60% may occur.

Starting temperature << ambient temperature: Use helium.
Starting temperature ≈ ambient temperature: Use nitrogen.

Use the same gas for calibration and experiment.

The importance of an inert atmosphere for correctly determining, for example the heat of fusion, is shown in Fig. 1.33. In the absence of an inert gas, some plastics decompose as they melt, with the extent of decomposition depending on the degree of stabilization. In such cases, it is not possible to identify a clear baseline in the molten state and quantitative determination of the enthalpy is difficult.

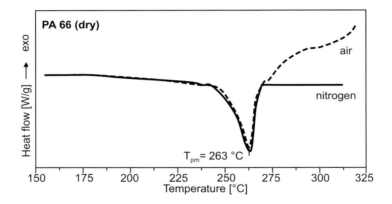

Fig. 1.33   Influence of nitrogen and air purge gas on the melting point curves of PA 66

T_{pm} = Melting peak temperature, 1st heating scan, specimen mass approx. 3 mg, heating rate 10 °C/min

1.2.2.5   Measuring Program

In **isothermal experiments**, it is necessary to minimize the transition while the specimen is equilibrating at the constant temperature. This is achieved by placing the specimen in the chamber at ambient temperature and heating it as quickly as possible to the desired temperature. At high heating rates, this may lead to brief "overshooting" of the temperature. Alternately, the specimen is heated to the isothermal temperature and then placed in the chamber, in which case the result is marked "undershooting".

In **temperature-scanning experiments**, the heating rate ranks next in importance to the starting and end temperatures. The **starting temperature** should be at least 50 °C beneath the temperature range of the expected transition because of the occurrence of undershooting.

This produces a stable baseline that permits reliable evaluation. Figure 1.34 illustrates such undershooting, which can extend for 10–20 °C, during measurement the glass transition of PET. At the same time, the very extent of the undershooting (1 mW in the diagram) shows the dimensions that it can assume relative to the glass transition of the PET.

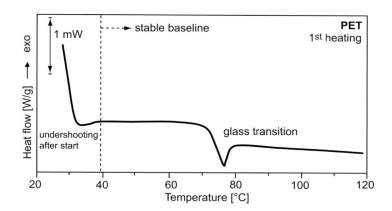

Fig. 1.34   Undershooting after the start of a DSC experiment performed on PET

1ˢᵗ heating scan, specimen mass approx. 3 mg, heating rate 10 °C/min, purge gas: nitrogen

Class of polymer	Starting temperature
Thermoplastics*	Ambient temperature*
Elastomers	–150 to –80 °C
Cured thermosets	Ambient temperature
Uncured thermosets	–50 °C

Plasticized or elastomer-modified polymers require much lower starting temperatures.

The practical consequence for studying melting in semicrystalline thermoplastics, which usually melt far above 100 °C, is to start the experiment at ambient temperature. The same applies to the standard amorphous thermoplastics.

Measurements of most cured thermosets may also be started at ambient temperature. Enthalpy determinations of cold-curing resin systems require even lower starting temperatures because they start reacting at ambient temperature.

Very low starting temperatures of down to -150 °C are required for determinations of high plasticizer fractions (e.g., in PVC), elastomeric components (butadiene in ABS), thermoplastic elastomers, or pure elastomers, because their glass transition temperatures are very low.

Prior to the start of the measurement, it must be ensured that the specimen has actually reached the temperature displayed. This is achieved by imposing an equilibration time of approx. 2 min for ambient temperature and of 10–20 min for lower temperatures. For useful evaluation of glass transitions, the **end temperature** should be approx. 50 °C higher than the transition. In melting determinations, the end temperature should be high enough to eliminate processing influences but not to cause thermal decomposition. In practice, the end temperature is about 30 °C higher than the final melting temperature T_{fm}.

Starting temperature:	**50 °C lower than the temperature range of expected transition,**
End temperature:	**30 °C higher, but risk of incipient decomposition**

The **heating rate** determines not only the duration of the experiment but also the results. It too must be adjusted to the transition being measured. The following heating rates are recommended in [4]:

Thermal transition	Heating rate
Glass transition	20 °C/min
Melting, crystallizing	10 °C/min*

Choose a higher heating rate to avoid any postcrystallization.

Other heating rates are possible, of course. Because the measured heat flux is directly proportional to the heating rate, higher heating rates generate a larger signal and can therefore serve to amplify particularly small transitions, for example, poorly defined glass transitions in highly filled thermosets or low content of crystalline impurities. The drawback is that a high heating rate leads to poorer resolution of adjacent melting peaks.

High heating rate: Amplifies small transitions, but impairs resolution.

Influence of Heating and Cooling Rates on the Glass Transition

Figure 1.35 shows how the heating rate affects the glass transition of polycarbonate, an **amorphous polymer**. As the rate increases, the step of the glass transition becomes larger and the glass transition becomes easier to evaluate.

The input thermal energy eliminates unstable structural changes (molecular orientation, internal stress, domains that have not yet crystallized, etc.) that were incorporated prior to the experiment, usually by processing, and adapts them, along with the equilibrium structures, to the current temperature as long as the transitions occur faster than the duration of heat input. This is the case above T_g. At the same time, c_p increases stepwise, and then remains perfectly constant as the temperature rises. At higher heating rates, the effect of the heat input is delayed until elevated temperatures are reached that trigger the transition of the more mobile macromolecules. This leads to a sudden jump in c_p and overshooting in the form of a local maximum. Accordingly, T_{mg} is displaced slightly toward higher temperatures.

Irrespective of the heating rate, the c_p value attained after the glass transition has been passed is of the same magnitude as it would otherwise have been. In Fig. 1.35, it is the direct analyse signal, that is, heat flux and not c_p that is plotted as a function of temperature. The c_p curves can be determined by dividing the heat flux by the heating rate (after calibration).

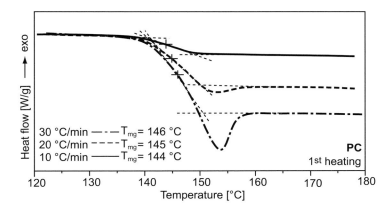

Fig. 1.35   Influence of heating rate on the formation of the glass transition of PC

T_{mg} = Midpoint temperature, 1st heating scan, specimen mass approx. 13 mg, purge gas: nitrogen

Increasing the heating rate enlarges the glass transition step, and displaces T_{mg} toward slightly higher temperatures.

Figure 1.36 shows the influence of the cooling rate on the glass transition in the 2nd heating scan, again for polycarbonate. The lower the cooling rate v_c, the more pronounced is the local maximum.

The shape of the glass transition is influenced by the preceding cooling rate.

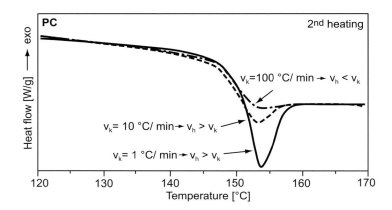

Fig. 1.36   Influence of cooling rate on the formation of the glass transition during the 2nd heating
              scan of PC

              v_c = Cooling rate, v_h = Heating rate,
              specimen mass approx. 11 mg, heating rate 20 °C/min, purge gas: nitrogen

Influence of Heating and Cooling Rate on the Melting and Crystallization Curves

DIN 53 765 recommends a heating rate of 10 °C/min for the melting of semicrystalline plastics. The heating rate greatly affects the experimental result, especially in specimens that tend to undergo postcrystallization or recrystallization. Because these are time-dependent processes, they are promoted by low heating rates.

Semicrystalline plastics have a more or less metastable structure because crystallization normally does not go to completion. Postcrystallization/recrystallization during heating may convert this metastable structure into a more stable form. In this case, the melting point curves do not reflect the material's original condition. Superposition of exothermic postcrystallization/recrystallization and endothermic melting may in extreme cases cause the melting peaks to virtually double in size. Postcrystallization is the conversion of amorphous domains into crystalline domains. Recrystallization, by contrast, is the conversion of one crystalline structure into another. The propensity for postcrystallization or recrystallization depends on the material and especially on its thermal history. Rapidly cooled or quenched specimens have a greater tendency to crystallize than postcured, more extensively crystallized specimens. For POM, a heating rate of 10 °C/min will likely permit the original state to be characterized, without the effect of post-crystallization/recrystallization. For PP, on the other hand, the same rate results in a marked change in enthalpy values, Fig. 1.37.

Low heating rates facilitate:

**Postcrystallization (amorphous → crystalline) and/or
Recrystallization (crystalline → crystalline, heat of fusion ΔH_m increases)
during measurement.**

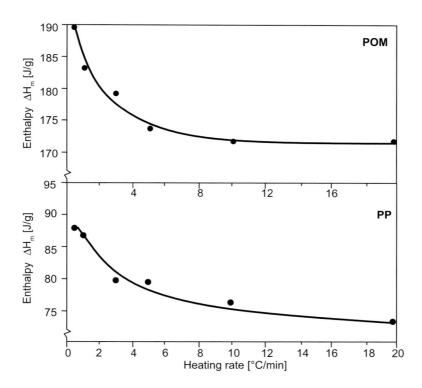

Fig. 1.37   Influence of heating rate on the heat of fusion ΔH_m of PP and POM, prior cooling rate $v_c = 10$ °C/min

Specimen mass approx. 3 mg, purge gas: nitrogen

With increasing heating rate, increasing thermal lag shifts the characteristic melting peak temperatures T_{pm} and especially the melting extrapolated onset temperature T_{eim} toward higher temperatures. At the same time, metastable structures, particularly at low heating rates, have sufficient time to convert into more perfect structures, for example crystallites with fewer defects and greater lamellar thickness, which then melt at higher temperatures (see Section 1.1.4.2).

Figure 1.38 is a plot of the melting peak temperature T_{pm} of PE-HD specimens with different histories as a function of the square root of the heating rate. Curves (b) and (c) are linear. Extrapolation to a theoretical heating rate of 0 yields a melting point of 132.3 °C for specimen (b) and 135.2 °C for specimen (c).

Curve (a) reflects typical postcrystallization during heating. When the heating rate is below 20 °C/min, recrystallization processes occurring during the measurement displace the melting peak toward higher temperatures. It is only when the heating rate exceeds 20 °C/min that recrystallization cannot occur within the timeframe of the experiment. At this stage, we are measuring the melting point, predetermined by the thermal lag of the apparatus, of the specimen's structure after quenching. Extrapolation of this part of the curve to a heating rate of 0 for the quenched specimen (a) returns a melting point of approx. 125 °C [15].

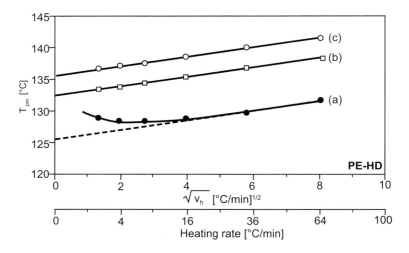

(a)   *1 mm thick slab quenched from 170 °C into ice water*
(b)   *Specimen (a) postcured for 50 h at 130 °C*
(c)   *Specimen (a) postcured for 167 h at 131.5 °C*

Fig. 1.38   Temperature of the melting peak of PE-HD specimens with different histories as a function of the square root of the heating rate [15]

T_{pm} = *Melting peak temperature, specimen mass 1 mg*

Propensity for recrystallization/postcrystallization depends on thermal history.

Cooling conditions during processing play a critical role in the structure of the polymer (see Section 1.1.4.3). In DSC, cooling conditions can be replicated by varying the **cooling rate**. Thus, the first heating curve allows conclusions to be drawn about prior processing. But it is also possible to create selected histories by choosing certain cooling rates.

Cooling conditions are responsible for the structure.

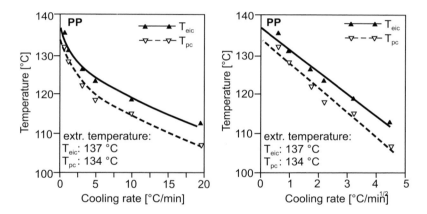

Fig. 1.39   Influence of cooling rate on the characteristic crystallization temperatures of PP;
               left: linear plot;   right: square root plot

               T_{eic} = *Crystallization extrapolated onset temperature,*
               T_{pc} = *Crystallization peak temperature,*
               End temperature of 1st heating scan at 200 °C,
               2nd heating scan, specimen mass approx. 3 mg, purge gas: nitrogen

High cooling rates shift the entire crystallization peak toward lower temperatures. Figure 1.39 shows how the cooling rate affects the characteristic temperatures of the crystallization curves of PP, the crystallization extrapolated onset temperature T_{eic} and the peak temperature T_{pc}. The temperatures have been plotted against the cooling rate (linear plot, left) and the square root of the cooling rate (right). Irrespective of the type of plot, extrapolation of the resultant curves yields the same values of 137 °C for T_{eic} and 134 °C for T_{pc}. As plotting the temperature against the square root of the cooling rate yields a linear relationship and makes extrapolation easier, this approach is preferable to the linear one.

High cooling rates displace the entire crystallization curve toward lower temperatures.

Because cooling rate directly influences crystallization behavior, uniform cooling rates should be used for comparative measurements. Figure 1.40 shows how different cooling rates affect melting of PBT. The specimen was cooled from the melt at different cooling rates (end temperature of 1st heating scan: 300 °C). Rapid cooling, at $v_c = 100$ °C/min (extensive supercooling), results in lower crystallinity and may lead to cold crystallization during heating. Slow cooling ($v_c = 1$ °C/min) gives a differently shaped melting point curve. Like other polymorphous polymers (e.g., polyamides), this specimen forms two different crystalline structures that have different melting temperatures (dual peak).

Use the same cooling rate to compare crystallization behavior.

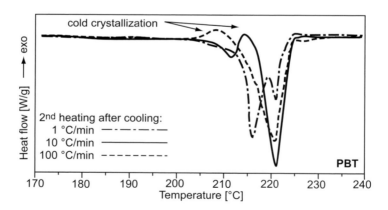

Fig. 1.40   Influence of cooling rate on the melting of PBT

Specimen mass approx. 5 mg, heating rate 10 °C/min, purge gas: nitrogen

Undefined, unknown cooling processes can cause considerable confusion when DSC heating curves are being interpreted. The holding time at the end temperature following the first heating scan, which itself is important, is also a major parameter that needs to be kept constant for comparative measurements. Figure 1.41 shows isothermal crystallization trials for PEK CF30 at a temperature of 349 °C. The material was probably not properly melted at a holding time of 1 minute at 420 °C, with the result that subsequent crystallization was random and nonreproducible. A holding time of 5 min seemed reasonable in this case, as it allowed reproducible values to be obtained.

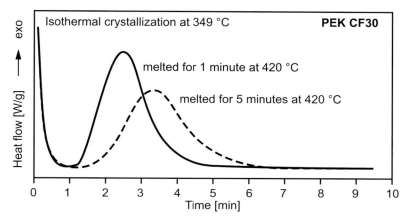

Fig. 1.41   Influence of holding time at the end temperature (in this case: 420 °C) on subsequent isothermal crystallization (in this case: 349 °C) of PEK CF30

Specimen mass approx. 10 mg, isothermal 349 °C/min, purge gas: nitrogen

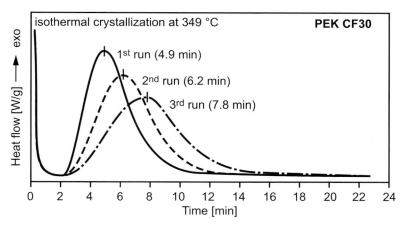

Fig. 1.42   Damage to a PEK CF30 through repeated use of a specimen, isothermal crystallization experiments at 349 °C

Specimen mass approx. 10 mg, isothermal 349 °C, purge gas: nitrogen

Measurements of isothermal crystallization processes can be very time-consuming and may be run overnight. However, the experiments must always be conducted on fresh specimens. Figure 1.42 shows a PEK CF30 that was used three times for isothermal crystallization under nitrogen at 349 °C. The crystallization peak becomes progressively narrower and shifts toward shorter times, which indicates that the material is damaged.

1.2.2.6   Evaluation

Evaluation of Glass Transition Temperature

While [1] and [4] are standardized methods for evaluating T_g from the midpoint temperature (equivalent to the half step height), software provided with the DSC instrument frequently permits other ways of determining it.

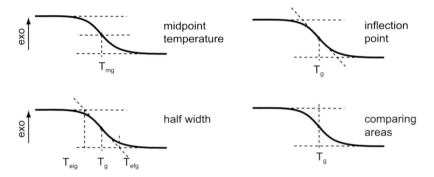

Fig. 1.43   Different ways of evaluating the glass transition step

If the glass transitions formed are ideal, that is, the baseline before and after the step is steady, all four evaluation methods yield the same result. However, unsteady baselines frequently make it difficult to apply a tangent. The **midpoint temperature** method (half step height) is a reliable evaluation method that is recommended both in current standards [1, 4] and in the literature [10, 14, 34].

Our experience bears out this approach: we conducted 10 measurements on each of PC (specimen mass: 10–14 mg) and EP resin (14–17 mg). The results are shown in Table 1.9. Evaluations based on the **half width** should yield the same   values. However, transitions that superpose the glass transition can hamper the determination of starting and end temperatures by this method.

As for the point of **inflection** results, the discrepancies are due to the fact that the software programs use different algorithms to determine it.

The method of **comparing areas** is little used in practice (but theoretically correct).

Power-compensation DSC

	Inflection point (T$_g$)* [°C]	T$_{mg}$ [°C]	Half width (T$_g$) [°C]
PC - mean value SD +/-	150 0.6	148 1.4	147 1.5
EP - mean value SD +/-	115 1.0	114 1.1	114 1.1

Heat-flux DSC

	Inflection point (T$_g$)* [°C]	T$_{mg}$ [°C]	Half width (T$_g$) [°C]
PC mean value SD +/-	151 2.3	149 1.0	149 1.3
EP mean value SD +/-	116 0.6	115 0.5	115 0.5

* In this case, the point of inflection had to be determined manually as the maximum of the first derivative.

Table 1.9   Influence of measuring technique and evaluation method on the experimental glass transition temperature of PC and EP, 2nd heating, SD = standard deviation

Specimen mass approx. 15 mg, heating rate 20 °C/min, purge gas: nitrogen

Different software algorithms yield different results.

These examples clearly show the good reproducibility of the experimentally determined glass transition temperatures. While the different evaluation methods return only minor differences for the two plastics employed, these may be much larger for curves of poor definition [35]. Comparison of the glass transition temperatures measured by two different methods (power-compensation DSC and heat-flux DSC) also shows good agreement between the values. In our experience, poorly defined glass transitions are much more readily identified and determined with the aid of a heat-flux calorimeter.

The user should always decide on one evaluation method and stay with it. Evaluating the glass transition can be problematic if it is superposed by an additional endothermic or exothermic effect. The inflection point method merits particular caution, because the edge used for the inflection point is higher and leads to higher values. In our experience, the midpoint method temperature method (half step height) also yields reliable results for poorly defined glass transitions.

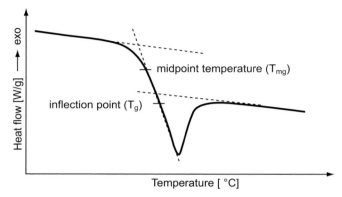

Fig. 1. 44 T_g evaluation, T_{mg} = midpoint temperature, T_g = point of inflection

Evaluate T_g as midpoint temperature T_{mg} (half step height).

Evaluation of Enthalpy

The evaluation of melting and crystallization curves has already been described in detail (see Section 1.1.4.2).

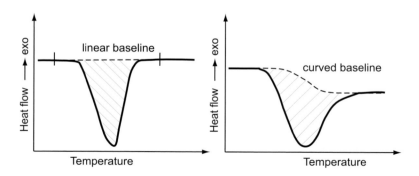

Fig. 1.45   Schematic drawing of different baseline types (linear and curved) for enthalpy integration

In enthalpy evaluations, usually the baselines before and after the transition are joined by a straight line, even though a change occurs in c_p during melting and crystallizing. In the case of chemical reactions that are accompanied by a marked c_p change, there is often not much point in drawing a straight connecting line. Cross-linking reactions in thermosets in particular can lead to considerable baseline shift. In such cases, the tangent before and after the transition is modified with the aid of a curved line.

Curved baselines are rarely used for the integration of melting or crystallization curves. They are more commonly employed for the evaporation of volatile substances or chemical reactions.

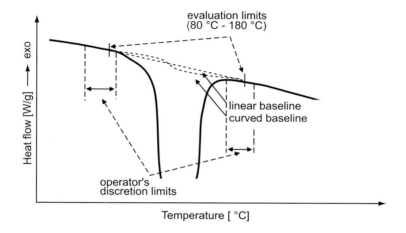

Fig. 1.46   Schematic diagram of a melting point curve with evaluation limits;
              linear and curved baseline for PP, as in Table 1.10

Not only is the slope different, but the temperatures at which the tangents for constructing the baseline are applied also affect the results considerably. An error-estimation experiment was performed on 10 specimens of PP pellets (specimen mass: 3–5 mg) with regard to melting peak temperature T_{pm} and heat of fusion ΔH_m. Different types of baseline (curved and linear) and temperature limits were employed, Fig. 1.46. The temperatures for applying the tangents were fixed at 80 °C and 180 °C in one case and were chosen individually at the operator's discretion in the other.

Construction of the baseline has no bearing on the melting peak temperature T_{pm}. The measuring principles are also comparable. In enthalpy evaluations, the experimentally determined value of ΔH_m depends heavily on the chosen baseline. If the curve is not ideally shaped, enthalpy differences of up to 30% (comparison of curved and linear baselines) can occur within the very same measurement. In summary, it may be said that the greater sensitivity of power-compensation DSC, relative to heat-flux DSC, also suffers from greater problems concerning the reproducibility of curve evaluation.

Power-compensation DSC

	Linear baseline Limits: 80 and 180 °C	Linear baseline Limits: operator's discretion	Curved baseline Limits: 80 and 180 °C
T_{pm} mean [°C] SD +/- [°C]	163 0.9	164 0.9	164 0.9
ΔH_m mean [J/g] SD +/- [J/g]	99.4 3.3	83.8 2.2	68.4 2.6

Heat-flux DSC

	Linear baseline Limits: 80 and 180 °C	Linear baseline Limits*: operator's discretion	Curved baseline* Limits: 80 and 180 °C
T_{pm} mean [°C] SD +/- [°C]	163 0.5	163 0.5	163 0.5
ΔH_m mean [J/g] SD +/- [J/g]	93.5 1.8	93.5 1.8	93.5 1.8

Due to a linear baseline, no difference was found between different evaluation limits in these cases. The swept baseline therefore becomes a linear line and does not return any differing values.

Table 1.10   Influence of analytical technique on the baseline shape and evaluation limits on melting peak temperature T_{pm} and heat of fusion ΔH_m , SD = standard deviation

Always use the same type of baseline
Choose evaluation limits judiciously after close inspection.

When making comparative measurements, it is advisable to fix the temperature limits or ranges. The greatest reproducibility was obtained through setting the limits in each case by inspection. Contamination or aging of the test chamber may give rise to unsteady, non-linear baselines. In this case, one possible remedy is to subtract a baseline produced by two empty pans from the experimental curve. That said, the goal must always to be to use a clean test chamber, this is, to clean it at regular intervals and always after chemical reactions.

1.2.3 Real-Life Examples

1.2.3.1   *Identification of Plastics*

Polymers have a characteristic molecular structure and morphology. DSC often enables unknown polymers to be identified from their characteristic melting and glass transition temperatures. Figure 1.47 shows melting peaks for different semicrystalline plastics, which can be reliably identified from the melting peak temperature T_{pm}. The curves stem from the 2nd heating scan, which was preceded by controlled cooling to eliminate thermal and processing history. It is clear from Figure 1.47 that the peak temperature may serve to distinguish between individual members of a particular class of polymer, for example, PA 12, PA 6 and PA 66 (see Table 1.3). The task of assigning peak temperatures to specific materials is simplified by using flame tests or density tests, to identify the polymer, that is, polyamide or polyolefin etc., in advance.

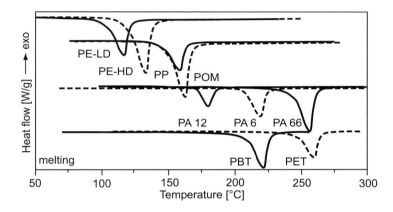

Fig. 1.47   Melting peaks of various semicrystalline thermoplastics

Specimen mass approx. 5 mg, heating rate 10 °C/min, purge gas: nitrogen

As the various examples in Fig. 1.48 show, amorphous thermoplastics have characteristic glass transition temperatures that generally permit unique identification. However, identification is more difficult when we are dealing with compatible polymer blends that yield a single glass transition temperature ranging between those of the two constituents (see Section 1.2.3.7). An example is PVC, whose glass transition temperature is depressed by added plasticizer. While rigid PVC has a glass transition temperature of approx. 80 °C, the T_g of plasticized PVC can be as low as -20 °C, the exact value depending on the content of plasticizer.

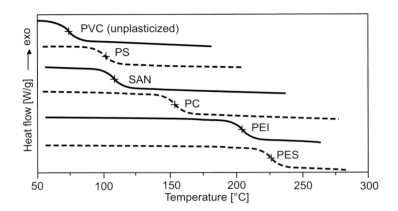

Fig. 1.48   Glass transitions of various amorphous thermoplastics

              Specimen mass approx. 15 mg, heating rate 20 °C/min, purge gas: nitrogen

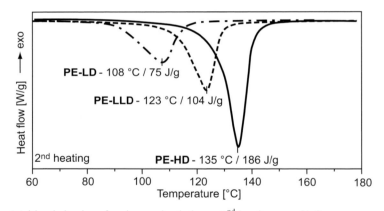

Fig. 1.49   Melting behavior of various polyethylenes, 2nd heating scan [36]

              Specimen mass approx. 5 mg, heating rate 10 °C/min, purge gas: nitrogen

The nature and form of monomeric starting materials determine the basic structure of
polymers, that is, whether they are linear, branched, or crosslinked. Bifunctional monomers
yield linear macromolecules (chain-like or thread-like molecules). Monomers with at least
three functions form branched or crosslinked structures or both.

In the case of polyethylene, for example, the degree of crosslinking is quantified by the
number of side-chains per 1000 atoms on the main chain:

PE-HD – Linear, with a few short chains (4–10 per 1000 C atoms),

PE-LD – Long chains to shrub-like branches,

PE-LLD– Linear, with many short chains (10–35 per 1000 C atoms).

The degree and type of branching give rise to substantial differences in properties (e.g., tendency to crystallize, hardness) [36]. As a result, differences occur in melting temperature and enthalpy, which can then be used to help identify the polymers.

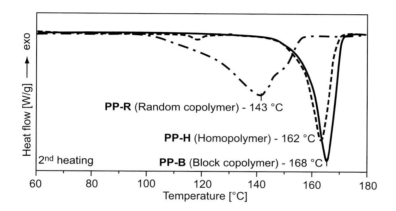

Fig. 1.50   2nd heating scans of different types of polypropylene

Specimen mass approx. 5 mg, heating rate 10 °C/min, purge gas: nitrogen

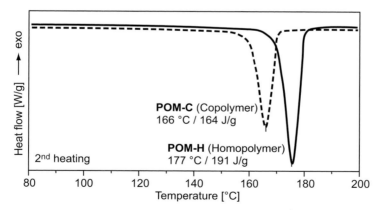

Fig. 1.51   Melting curves for POM homopolymer and copolymer, 2nd heating scans

Specimen mass approx. 5 mg, heating rate 10 °C/min, purge gas: nitrogen

A macromolecule whose building blocks or monomers are all the same is called a homo-polymer. If the monomers are different, the polymer is called a copolymer. There are many kinds of structural copolymers (e.g., random/statistical copolymers, sequential copolymers, segment copolymers, block or graft copolymers) [36]. Figure 1.50 sows the melting curves of different types of polypropylene [36].

A further example of the different melting behaviors of homopolymers and copolymers is shown in Fig. 1.51 for POM. The copolymer melts at a much lower temperature and is thus readily distinguished from the homopolymer.

1.2.3.2   Crystallinity

The properties of plastics are critically affected by their crystallinity (see Section 1.1.4.2). The more crystalline the molded part, the more rigid and stronger it is but the more brittle it is also. The crystallinity of a polymer is influenced by its chemical structure, cooling conditions during processing, and any thermal posttreatment.

Crystallinity is influenced by the chemical structure and thermal history.

Fig. 1.52 shows the melting of a molded POM part in the 1st heating curve. The crystallinity is determined by dividing the experimental heat of fusion ΔH_m = 173 J/g by the literature value for POM ΔH_m^0 = 326 J/g (see Section 1.1.4.2). For this specimen, the degree of crystallization computes to w_c = 53%.

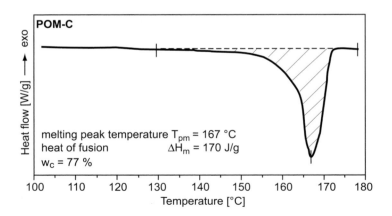

Fig. 1.52   Heating curve of POM

1st heating scan, specimen mass approx. 3 mg, heating rate 10 °C/min,

purge gas: nitrogen

$$w_c = \frac{\Delta H_m}{\Delta H_m^0} \cdot 100 = \frac{173 \quad J/g}{326 \quad J/g} \cdot 100 = 53 \quad \left[\%\right]$$

In certain circumstances, calculating the crystallinity can help identify a polymer. In the case under discussion, the peak temperature of 167 °C would indicate that the specimen is PP. When we calculate the crystallinity using the literature value of 207 J/g for PP, we obtain a value of 84%, a figure that is very high and unrealistic. The plastic is therefore more likely to be POM. Further information can be obtained by simple methods of determination, such as density, combustion, and so forth.

A low crystallinity may be the result of low mold wall temperatures and high cooling rates during processing, especially if the molded part has low section thickness. When such a part is used at elevated temperatures, shrinkage and other undesirable effects may occur due to postcrystallization. Postcrystallization can be deliberately induced by postcuring. Figure 1.53 shows how the crystallinity of injection molded parts increases with rise in mold wall temperature and various annealing conditions.

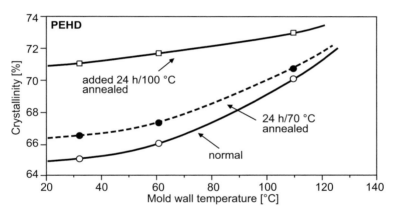

Fig. 1.53   Crystallinity of injection molded parts as a function of mold wall temperature and annealing [36]

1.2.3.3   Thermal History

In a DSC measurement, information about thermal and mechanical history is revealed by the 1st heating curve. The 2nd heating curve is used for determining material characteristics in comparative trials under the given dynamic conditions (end temperature of 1st heating scan, cooling rate, heating rate of 2nd heating scan).

Figure 1.54 shows the 1st heating scan for PE-HD, starting from ambient temperature and increasing to 200 °C at a rate of 10 °C/min. The specimen is then cooled at a controlled rate of 10 °C/min to ambient temperature (cooling curve) and reheated to 200 °C (2nd heating scan). In each case, the peak temperature and the fusion or crystallization enthalpy are evaluated. The 2nd curve has a higher peak temperature and higher heat of fusion, and thus greater crystallinity than the 1st curve. The heat of fusion ΔH_m of the 2nd heating curve roughly corresponds to the crystallization enthalpy ΔH_c from the cooling curve and is approx. 20% higher than the heat of fusion from the 1st run. This indicates that the cooling rate during processing was higher than in the DSC experiment.

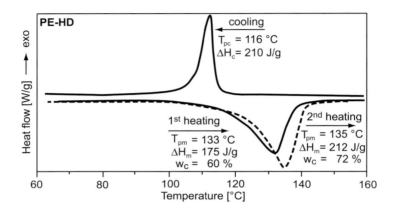

Fig. 1.54   1st heating scan, defined cooling and 2nd heating scan of a PE-HD specimen

Specimen mass approx. 3 mg, heating rate 10 °C/min, cooling rate 10 °C/min, purge gas: nitrogen

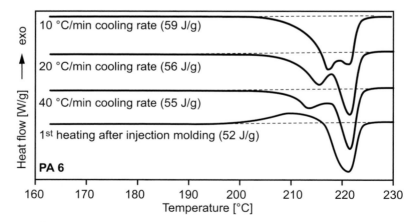

Fig. 1.55   2nd heating scan for a PA 6 that had been cooled at different rates

Specimen mass approx. 2 mg, heating rate 10 °C/min, purge gas: nitrogen

The influence of the cooling rate on the 2nd heating scan is again illustrated with PA 6 in Fig. 1.55 (see also Fig. 1.39). The specimens were each melted for 2 min at 280 °C and then cooled at different rates. This leads to two crystal modifications, the magnitude of which can provide an indication of the rate at which a specimen has been cooled.

The literature also reports two peaks for PA 6 at 215 °C and 223 °C and attributes them to a lower-melting γ-modification and a higher-melting α-modification. Given the melting behavior of the specimens that were made at different cooling rates, it is reasonable to assume that the dual peak is formed in a time-dependent recrystallization process.

1st heating curve: Thermal and mechanical history
2nd heating curve: Material characteristics following controlled cooling

Further evidence of thermal history is provided by **cold crystallization**, which occurs during heating (see Section 1.1.4.3). PET serves to illustrate this effect. Figure 1.56 shows the 1st heating curve of such a specimen that was cooled very rapidly from the melt to below T_g, a process that has largely suppressed crystallization. The result is a translucent, predominantly amorphous material such as that used in beverage bottles. The 1st heating curve features an endothermic glass transition step that is fairly high for a semicrystalline thermoplastic. From about 140 °C on, cold crystallization (ΔH_c = 23.9 J/g) sets in because of greater segment mobility. The newly formed and already existing crystallites start to melt at about 220 °C; the heat of fusion ΔH_m is 33.2 J/g. Comparison of the heat of fusion of 33.2 J/g with the crystallization enthalpy ΔH_c = 23.9 J/g shows that most of the crystallites have formed only during heating.

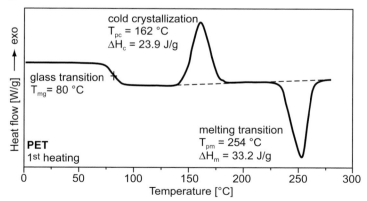

Fig. 1.56   Heating curve for PET showing the glass, cold crystallization and melting transitions

Specimen mass approx. 3 mg, heating rate 10 °C/min, purge gas: nitrogen

New ways are being devised of using TMDSC to determine this original crystallinity while taking into account the reversible heat of fusion [33, 42].

The effect of annealing on a PPS specimen is shown in Fig. 1.57. When a unannealed specimen is heated, the 1st heating curve shows signs of cold crystallization at approx. 110 °C. Subsequent **annealing** of the molded part, for 3 h at 200 °C in this example, increases the crystallinity. A small endothermic peak, known as the postcuring peak, is formed at approx. 210 °C. This is due to melting of the small, less thermostable crystallites formed at the postcuring temperature.

To prevent such a transition from being falsely interpreted, for example, as indicating a second component (perhaps PA 6), it is necessary to perform a 2nd heating scan. This will cause the postcuring peak to disappear. The reappearance of a melting peak would suggest that a second component was present.

> **Annealing below the melting peak temperature (T_{pm}) produces an additional endothermic peak.**

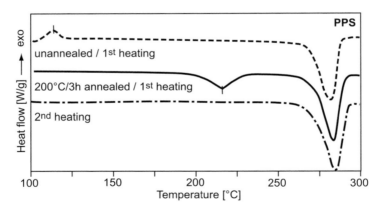

Fig. 1.57   Melting point curves for PPS specimens of different thermal history

1st and 2nd heating scan, specimen mass approx. 5 mg, heating rate 10 °C/min, purge gas: nitrogen

The annealing temperature determines the position of the annealing peak or a annealing "shoulder", Fig. 1.58 and 1.59. If it is above T_g, very small crystallites of different lamellar thickness are formed, whose exact size depends on the annealing temperature. They melt again at a temperature some 10–15 °C above the annealing temperature.

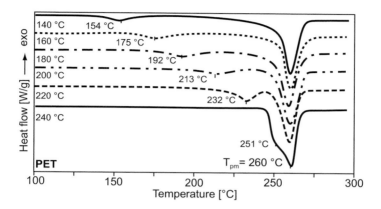

Fig. 1.58   Influence of different annealing temperatures on the melting curve of PET
               specimens; all annealed for 15 h

1ˢᵗ heating scan, specimen mass approx. 5 mg, heating rate 10 °C/min,

purge gas: nitrogen

If annealing is performed close to the melting temperature, a so-called annealing shoulder
forms that cannot be completely resolved from the melting peak. Figure 1.59 shows typical
annealing shoulders for PE-HD. As the annealing temperature rises, not only does the
position of the shoulder increase but also its magnitude. This is due to growth in the number
of crystallites of the same lamellar thickness.

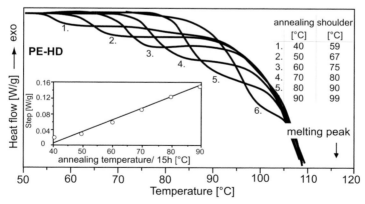

Fig. 1.59   Influence of annealing temperature on the melting curve
               (inset: initial area magnified) of PE-HD specimens (all postcured for 15 h)

1ˢᵗ heating scan, specimen mass approx. 5 mg, heating rate 10 °C/min,

purge gas: nitrogen

**The temperature position of a annealing peak or a postcuring shoulder depends
on the annealing temperature.**

The actual service temperatures of plastics can be reproduced and assessed by performing DSC annealing trials at specific temperatures and for specific periods of time. The practical conditions must be reproduced as faithfully as possible, for example, the entire molded part should be annealed, perhaps in a certain medium, or samples must be removed from the same position from specimens to be compared.

For several plastics, it is common to employ a staggered annealing program so as to optimize the property profile by influencing the crystallinity. PEEK was annealed at five temperatures, starting from 340 °C (340, 330, 320, 310 and 300 °C, Fig. 1.60). The subsequent heating curve has a annealing peak for each of the five annealing stages. Overall, the crystallinity of the specimen increases by roughly 25% relative to that of the unannealed material.

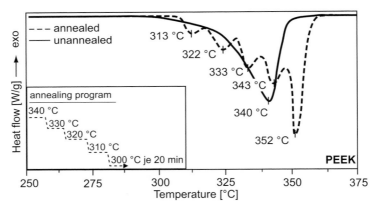

Fig. 1.60   Influence of a annealing program on the melting point curve of PEEK relative to
              unannealed material [47]

              1st heating scan, specimen mass approx. 7 mg, heating rate 20 °C/min,

              purge gas: nitrogen

The annealing time is particularly influential at high postcuring temperatures. Figure 1.61 shows the influence of the postcuring time on peak shape and the temperature position of the DSC curve for an unfilled PA 6. All specimens were postcured at 210 °C.

As the annealing time increases, from several seconds to 10 min, the start of melting shifts to a temperature approx. 5 °C higher, the enthalpy increases slightly and there is a dramatic change in the shape of the curve. The crystal modifications characterized by a dual peak are always further transformed into one crystal modification, Fig. 1.55. The end of the melting process is not affected.

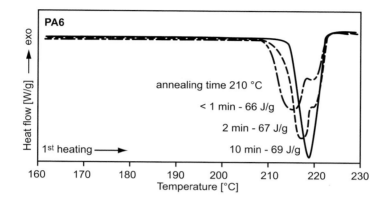

Fig. 1.61   Melting curve for a PA 6 following different annealing times at 210 °C

1ˢᵗ heating scan, specimen mass approx. 3 mg, heating rate 10 °C/min,
purge gas: nitrogen

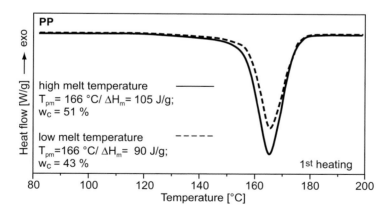

Fig. 1.62   Influence of melt temperature on the melting behavior of a PP part
              during injection molding

T_{pm} = *Melting peak temperature, ΔH_m = Heat of fusion,*
1ˢᵗ heating scan, specimen mass approx. 4 mg, heating rate 10 °C/min,
purge gas: nitrogen

The 1st heating scan of a DSC experiment also yields clues about the **processing conditions**. It is not possible to directly identify crystallinity, melt temperature, or mold temperature. Nevertheless, comparisons can be made and hence tendencies quoted. Whether PP is injection molded at a melt temperature of 210 or 270 °C makes quite a difference to the crystallinity of the molded part (Fig. 1.62).

In both cases, the mold temperature was 40 °C. The melting point curves show identical peak temperatures, but the PP processed at the higher melt temperature is more crystalline because it had more time to crystallize.

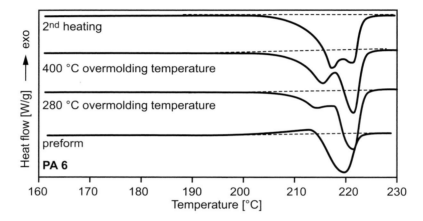

Fig. 1.63   Melting curves for PA 6 specimens from surface regions of the two-component
                 preform before and after overmolding with the second component at different
                 melt temperatures, along with 2nd heating scan

                 1st heating scan, specimen mass approx. 3 mg, heating rate 10 °C/min,
                 purge gas: nitrogen

In two-component injection molding, a preform is overmolded with a second component. The influence of the melt temperature of the second component on the layer close to the surface of the first component is shown in Fig. 1.63.

The 1st heating scan for the PA 6 preform is a typical curve that indicates slight re-crystallization and a subsequent melting peak. The overmolding temperatures of 280 °C and 400 °C for the second component   lead to the formation of two peaks that become more distinct as the temperature rises. For comparison, the 2nd heating scan is also shown in Fig. 1.63. The dual peak produced by the PA 6 under the aforementioned cooling conditions can be seen.

> **Annealing effects are no longer recognizable once the thermal history has been eliminated in the 2ⁿᵈ heating scan.**

1.2.3.4   Water Uptake

Plastics particularly polar ones like polyamides absorb water from their surroundings. Whether accidental or deliberate (e.g., through conditioning), absorption of water affects their properties (see Section 1.2.2.1). Water acts as a plasticizer in the polymer and thus lowers the glass transition temperature. Figure 1.64 shows how water affects the 1ˢᵗ heating curves of amorphous PA.

> **T_g of polyamides falls considerably with rise in water content.**

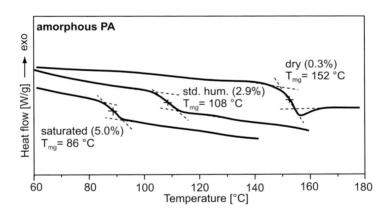

Fig. 1.64   Influence of water content (dry, standard human, saturated) on the glass transition of amorphous PA

1ˢᵗ heating scan, specimen mass approx. 14 mg, heating rate 20 °C/min,

purge gas: nitrogen

Because it is mainly the amorphous domains that absorb water, the extent of absorption by semicrystalline polyamides varies with the crystallinity. The saturation concentration of water and the rate of water absorption by polyamides depend not only on the crystallinity but also on the number of polar amide groups and thus on the type of polyamide.

Figure 1.65 shows how water absorption varies with PA structure, that is, the ratio of CH_2 groups to NHCO (amide) groups.

The physical properties of polyamides can be assessed only if their water content is known. This is why characteristic values are usually specified for three different moisture levels (equilibrium conditions) [39]:

dry = ≈ *0% rel. humidity (e.g., freshly molded)*
standard human = *Stored at 23 °C/50% rel. humidity to constant weight*
saturated = *Stored at 100% rel. humidity (immersion in water) to constant weight*

Thermoplastics with a water content of 0–0.2% are termed dry. Those containing 1.5–2.7% water are called air dry and those with 5–8% water are said to be wet. Because these percentages are percentage by weight, the presence of fillers that do not absorb moisture must be borne in mind.

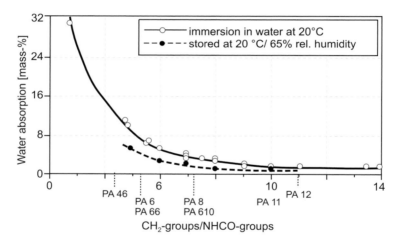

Fig. 1.65   Water absorption by different polyamides as a function of chemical structure [39]

1.2.3.5   Nucleation

Nucleating agents, contaminants, and processing can all influence nucleation and thus crystallization. The number of crystal nuclei present determines the number and size of crystalline superstructures formed. Nucleating agents (short-chained polymers) generate more nuclei and, under the same cooling conditions, a more finely grained, spherulitic structure.

Figure 1.66 shows the effect of deliberately added nucleating agents and of their concentration on the crystallization of PP. As the concentration of nucleating agents increases, the whole curve shifts toward higher temperatures and the start of crystallization T_{eic} increases by roughly 6 °C. The crystallization curve is essentially characterized by T_{eic} and T_{pc}. The tail of the crystallization curve flattens out toward low temperatures and so the end of crystallization cannot be completely detected by DSC. T_{efc} is taken to be the end point.

Conc. of nucleating agent	T_{eic} [°C]	T_{pc} [°C]	T_{efc} [°C]
0.001%	119	115	111
0.01%	122	118	115
0.05%	124	120	117
0.1%	125	121	118

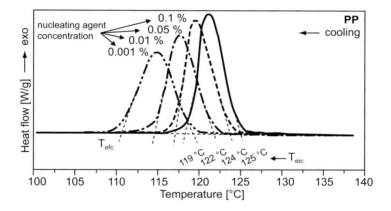

Fig. 1.66   Effect of nucleating agent concentration on the crystallization of PP

T_{eic} = Crystallization extrapolated onset temperature,

T_{pc} = Crystallization peak temperature,

T_{efc} = Crystallization extrapolated onset temperature

Cooling curve, specimen mass approx. 3 mg, cooling rate 10 °C/min, nitrogen

The influence of nucleating agent concentration on the 2nd heating scan is shown in Fig. 1.67. Specimens with a low concentration crystallize at lower temperatures and therefore tend to recrystallize in the form of a dual peak during heating and melting at the given heating rate. Although a greater concentration reduces the spherulite size in a finer structure, the lamellae that make up the spherulites become thicker on account of the higher crystallization temperature.

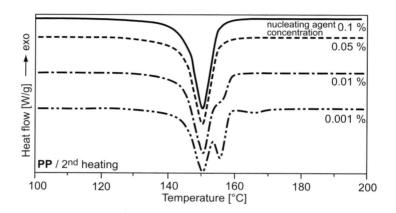

Fig. 1.67   Effect of nucleating agent concentration on the melting of PP

Specimen mass approx. 3 mg, heating rate 10 °C/min, purge gas: nitrogen

Addition of nucleating agents
Crystallization curve shifted to higher temperatures
Modified crystallization influences 2nd heating curve.

1.2.3.6   Aging

DIN 50035 Part 1 [40] classifies plastics aging as being either physical or chemical. Physical aging changes the morphology of plastics (postcrystallization, crystal structure, orientation, internal stress, etc.) while chemical aging changes their chemical structure (cleavage of the polymer chains, crosslinking, oxidation).

Physical aging can lead to effects similar to those induced by annealing at high temperature, such as the formation of peaks or a rise in peak temperature and heat of fusion. This can be clearly seen in the heating curve for a   PP specimen that was aged for 8 h at 160 °C, Fig. 1.68.

Annealing at a temperature in the middle of the melting range causes small crystallites with thin lamellae to melt and recrystallize to form more perfect crystals having thicker lamellae. The melting range of this specimen thus shifts toward higher temperatures relative to that of the non-annealed specimen. The annealing temperature can be estimated from a shoulder in the melting point curve and from the shift of the melting peak.

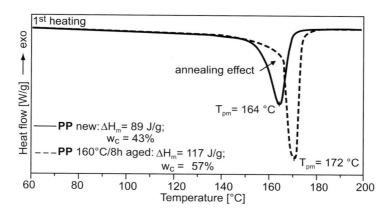

Fig. 1.68   Influence of physical aging on the melting of virgin and annealing PP, 8 h at 160 °C

T_{pm} = Melting peak temperature, ΔH_m = Heat of fusion,
w_c = Crystallinity (expressed in terms of ΔH_m^0 = 207 J/g),
1st heating scan, specimen mass approx. 5 mg, heating rate 10 °C/min,
purge gas: nitrogen

Subsequent cooling and the 2nd heating curve (not shown here) fail to reveal any differences in the crystallization and melting profiles of the two specimens.

It may therefore be assumed that the chemical structure of the material is unaffected by this physical aging.

Evidence of physical aging revealed by

1st **Heating curve: Higher melting peak temperature and crystallinity, annealing effect**

Cooling curve, 2nd Heating curve: None

Figures 1.69 and 1.70 show the changes in chemical structure that have occurred in a molded PP part used for several years at roughly 80 °C. The crystallization and melting behavior are different than those of a new part.

The crystallization curve of the aged molded part has two crystallization peaks, whereas the new part has only one, Fig. 1.69.

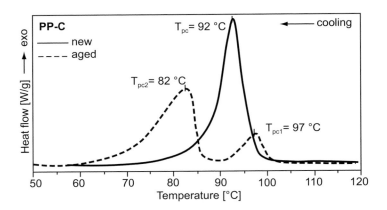

Fig. 1.69   Cooling curves for PP copolymer; new and aged for several years at approx. 80 °C

T_{pc} = Crystallization peak temperature, specimen mass approx. 5 mg, cooling rate 10 °C/min, purge gas: nitrogen

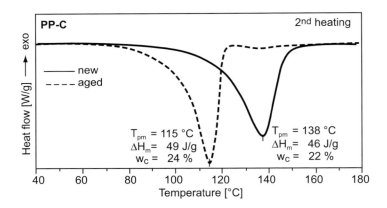

Fig. 1.70   Chemical aging of PP copolymer; new and aged for several years at roughly 80 °C [41]

T_{pm} = Melting peak temperature, ΔH_m = Heat of fusion, w_c = Crystallinity, 2nd heating, specimen mass approx. 5 mg, heating rate 10 °C/min, purge gas: nitrogen

The 2nd heating curves recorded afterwards show a much greater drop in the peak temperature of the aged part than in the new part. In the absence of a reference material, it would be
concluded that the material was PE. Infrared measurements can show, however, that the

material is PP. The depressed peak temperature is due to permanent structural change (molecular chain degradation, etc.).

Note: *DSC is particularly good at identifying chemical aging in polyolefin's. It is less suitable for other plastics, for example, polyamides, because chemical aging of them does not necessarily manifest itself as a change in crystallinity or peak temperature.*

Evidence of chemical aging revealed by

1st Heating curve: **Influence of peak temperature and crystallinity, perhaps superposed by physical aging**

Cooling curve: **Nucleation effect**

2nd Heating curve: **Reduced crystallinity and perhaps lower peak temperature**

Chemical aging in amorphous thermoplastics is indicated by a lowering of the glass transition temperature, as shown in Fig. 1.71 for PMMA.

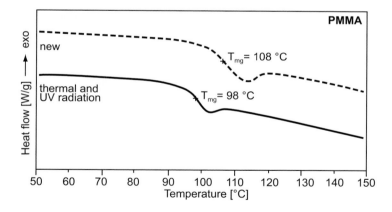

Fig. 1.71   Influence of UV radiation and heat on PMMA

2nd heating scan, specimen mass 15 mg, heating rate 20 °C/min, purge gas: nitrogen

The PMMA was used for the canopy of a street lamp that was exposed to both thermal and UV radiation. The lowering of the glass transition temperature is probably due to plastification by degraded components (plasticizers).

Chemical aging of amorphous thermoplastics: Lowering of T_g

1.2.3.7   Crosslinking of Thermoplastics

Thermoplastics can also be made to crosslink in three dimensions so that certain material properties may be improved. A common process is electron-beam crosslinking, which allows the molded part to be crosslinked to various depths, the extent of which depends on the radiation source (α- or β-emitter). This leads to a shift in glass transition temperature and a change in melting and crystallization behavior.

Figure 1.72 shows the effect of electron-beam treatment on the melting curves of a molded PA 6 part that was irradiated with 99 kGy. The peak temperature drops by approx. 15 °C and the heat of fusion by 20–30%. These effects can be seen in both the 1ˢᵗ and 2ⁿᵈ heating scans. In other words, a chemical change is occurring in the material.

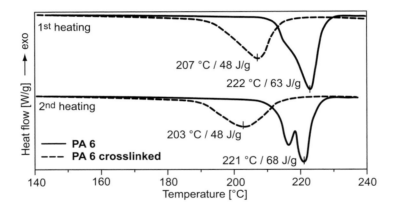

Fig. 1.72   Influence of electron-beam crosslinking on the melting of PA 6

1ˢᵗ and 2ⁿᵈ heating scans, specimen mass approx. 3 mg, heating rate 10 °C/min,
purge gas: nitrogen

There is also a substantial change in the cooling curves. The corresponding influence of electron-beam treatment of PBT is shown in Fig. 1.73. The crystallization peak of the crosslinked material shifts completely to lower temperatures; in this example, crystallization now begins 23 °C earlier.

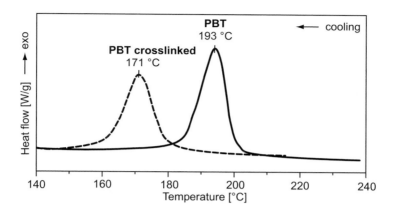

Fig. 1.73   Influence of electron-beam treatment on the crystallization of PBT

Cooling, specimen mass 3 mg, cooling rate 10 °C/min, purge gas: nitrogen

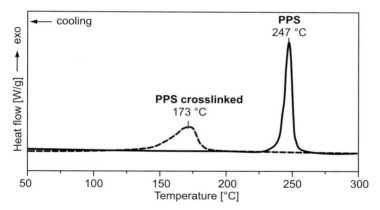

Fig. 1.74   Influence of thermal crosslinking, relative to previous end temperature,
             on the crystallization of linear PPS

Cooling, specimen mass 3 mg, cooling rate 10 °C/min, purge gas: nitrogen

It is possible in such cases to wrongly identify materials with the aid of DSC. Aside from using other techniques, such as infrared spectroscopy, it is always advisable to analyze the inside of the molded part, as selective crosslinking occurs only at the surface.

The example above illustrates the effect of thermal crosslinking of a thermoplastic. This was a linear PPS that crosslinks between 380 °C and 440 °C, as established in a previous experiment.

Figure 1.74 shows the difference in crystallization behavior. The material was heated to 320 °C in the 1ˢᵗ heating scan and then, for comparison, to 450 °C. The crystallization temperatures have shifted by more than 60 °C and the heat of crystallization by approx. 50%. The 2ⁿᵈ heating scan curves measured after defined cooling show a roughly 16 °C shift in the peak and a change in the heat of fusion of almost 40%.

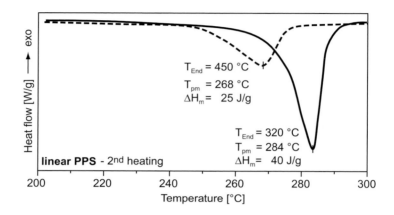

Fig. 1.75   Influence of heat treatment on the melting behavior of linear PPS,
               T_{End} = end temperature of the 1ˢᵗ heating scan

               2ⁿᵈ heating scan, specimen mass 15 mg, heating rate 20 °C/min, purge gas: nitrogen

1.2.3.8   Mixtures, Blends, and Contaminants

Thermoplastic polymers are blended selectively to boost certain properties. These may be enhanced processability, paintability, impact modification, increased thermal resistance and a reduction in susceptibility to stress cracking. In contrast, contaminants that get into the molded part via processing or via dirty raw material can impair its properties or adversely affect processing.

DSC studies permit not only identification of certain thermoplastic components but also a quantitative estimation of the blending fractions, the extent of this depending on the morphology.

Plastics generally have limited miscibility owing to energetic reasons and so it is rare to find blends that are homogeneous at the molecular level. Demixing is usually due to in-compatibility.

Amorphous/Amorphous

Two compatible amorphous polymers ideally have a common glass transition in between those of the individual constituents. Incompatible mixtures are much more common. Each constituent yields a glass transition at its original temperature, that is, the properties of each constituent are more or less retained in the mixture. Figure 1.76 shows a typical example of an incompatible mixture of ABS and PC.

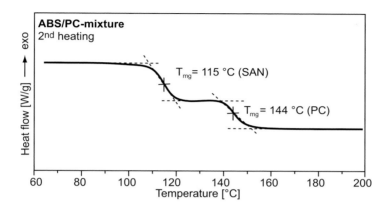

Fig. 1.76   Glass transition of PC and SAN (ABS constituent) in an ABS/PC mixture

2nd Heating scan, specimen mass approx. 15 mg, heating rate 20°C/min, purge gas: nitrogen

Compatible amorphous mixtures: Common glass transition
Incompatible amorphous mixtures: Two glass transitions

The relative step heights of the two glass transitions of PC and the SAN contained in the ABS can be used to quantitatively estimate the ABS and PC fractions. This is illustrated in the calibration curve below (Fig. 1.77), which is a plot of the step height produced by ABS mixtures of different, known compositions. As the ABS fraction increases, the step height of the SAN increases almost linearly. The SAN content in a specimen of unknown composition can therefore be determined in this concentration range (80–100% ABS).

Other examples of incompatible amorphous/amorphous mixtures are:
  − PC mixed with PMMA to increase UV resistance,
  − PMMA mixed with ABS to increase the weatherability and rigidity of ABS.

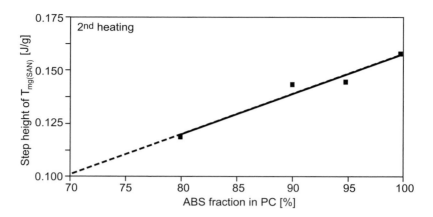

Fig. 1.77   Step height of the SAN glass transition in ABS/PC calibration mixtures

*2nd Heating scan, specimen mass approx. 15 mg, heating rate 20°C/min,
purge gas: nitrogen*

Amorphous/Semicrystalline

In certain circumstances, amorphous/semicrystalline mixtures can behave homogeneously in
their glass transition ranges. This is the case, for example, for mixtures of amorphous PEI
and semicrystalline PEKEKK, which have a common glass transition temperature. Figure
1.78 illustrates this for different mixture compositions.

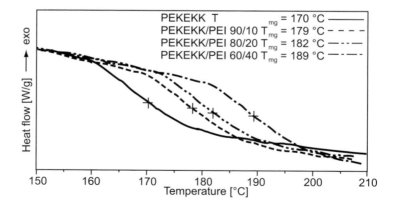

Fig. 1.78   Glass transitions of compatible PEKEKK/PEI mixtures [47]

*1st Heating scan, specimen mass approx. 5 mg, heating rate 20 °C/min,
purge gas: nitrogen*

The midpoint temperature T_{mg} of this common glass transition changes linearly with the mass fractions of the components, Fig. 1.79. This calibration curve can be used to estimate the mixing ratio of PEI/PEKEKK mixtures of unknown composition.

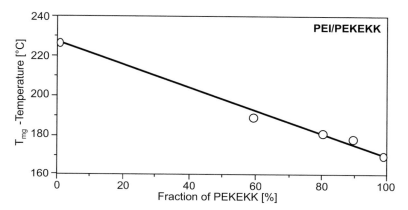

Fig. 1.79   Glass transition temperature T_{mg} as a function of PEKEKK mass fraction in PEI

If the melting peak and the glass transition do not overlap, the semicrystalline polymer in amorphous materials can also be identified from its melting peak. Measurements performed on mixtures of ABS with PBT, POM, and PP in concentrations ranging from 5–20% have shown that the heat of fusion varies directly with the fraction of semicrystalline constituent (Fig. 1.80).

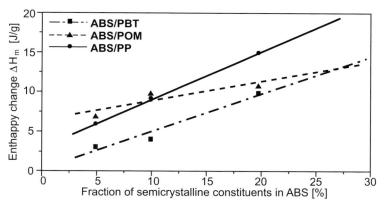

Fig. 1.80   Change in heat of fusion of semicrystalline constituents PBT, POM and PP mixed with ABS as a function of content over the range 5–20%

2nd Heating scan, specimen mass approx. 10 mg, heating rate 20 °C/min,

purge gas: nitrogen

Extrapolation should not be performed beyond the experimental standards as the heat of fusion and the mixing ratio, due to mutual effects on crystallization, do not behave linearly across the entire range. Nevertheless, small fractions (up to 0.5%) of semicrystalline polymer can still be identified.

Further examples of mixtures of amorphous and semicrystalline constituents are PPE/PA, PPE/PBT, and PP/EP(D)M.

Semicrystalline/Semicrystalline

Mixtures of semicrystalline constituents are also used for optimizing certain properties. Typical examples are PP/PE and PA/PE, with the PE content intended to enhance low-temperature impact strength.

Mostly, DSC heating curves for mixtures of two semicrystalline constituents also contain two melting peaks, that is, the mixtures are heterogeneous and incompatible. Because the two constituents can also affect each other's crystallization and melting behavior (temperature position and peak area), quantitative interpretation of their heating curves is often difficult.

> **Quantitative determination by means of heat of fusion is difficult.**
> **That is perhaps possible through use of calibration curves for the same mixing partners and known compositions.**

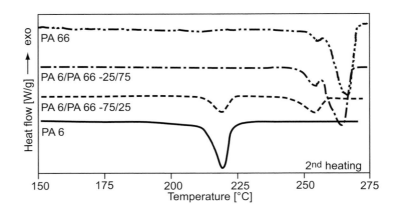

Fig. 1.81   Melting point curves for PA 6/PA 66 mixtures

2nd Heating scan after cooling at 10 °C/min, specimen mass approx. 5 mg, heating rate 10 °C/min, purge gas: nitrogen

Figure 1.81 illustrates this for mixtures of PA 6 and PA 66. 25 parts by weight PA 6 are not detectable in PA 66 but, in the reverse situation (25 parts by weight PA 66), there is a melting peak for both PA 6 (at 220 °C) and PA 66 (at 265 °C).

Note: *The dual peak is due to two different crystal modifications found also, for instance, in PBT (Fig. 1.37).*

The cooling curves for 25 pbw PA 6 in PA 66 and for 25 pbw PA 66 in PA 6 (Fig. 1.82) contain only one crystallization peak. The crystallization peak for 25 pbw PA 6 is almost in the same position as that for pure PA 66. In contrast, the crystallization peak for 25 pbw PA 66 occurs between those for pure PA 6 and PA 66.

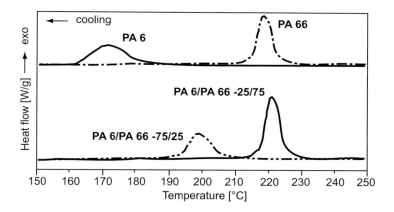

Fig. 1.82   Crystallization curves for PA 6/PA 66 mixtures

Specimen mass approx. 5 mg, cooling rate 10 °C/min, purge gas: nitrogen

An example of the behavior of two semicrystalline materials having little structural similarity is shown in Fig. 1.83. This is the heating curve for PA 6, which was treated with PE to improve the impact strength at low temperatures. Aside from the dominant melting peak of the PA 6 (at T_{pm} 224 °C), the 1st heating curve contains a small melting peak due to PE (at T_{pm} 125 °C).

The cooling curve clearly reveals the influence of the PE on the processing of PA 6 (Fig. 1.84). The added polyethylene acts as a nucleating agent and displaces the crystallization curve (i.e., the start of crystallization) by 5 °C toward higher temperatures.

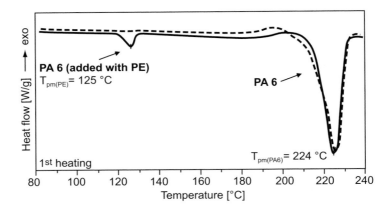

Fig. 1.83   Melting point curves for PA 6 pellets, with and without added PE

T_{pm} = *Melting peak temperature, 1ˢᵗ heating scan, specimen mass approx. 3 mg,
heating rate 10 °C/min, purge gas: nitrogen*

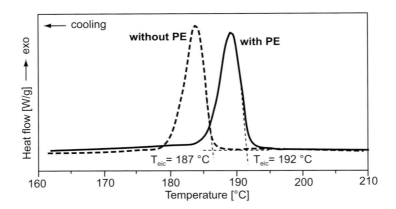

Fig. 1.84   Crystallization of PA 6 pellets with and without added PE

T_{eic} = *Crystallization extrapolated onset temperature, specimen mass approx. 3 mg,
cooling rate 10 °C/min, purge gas: nitrogen*

Changes in crystallinity can also adversely affect part quality, for example, if they are caused by contaminants. A flawed, molded PA 6 part that suffered brittle fracture after injection molding was found to have a small shoulder at approx. 230 °C at the end of its melting peak. The undamaged reference part showed no such anomaly, Fig. 1.85.

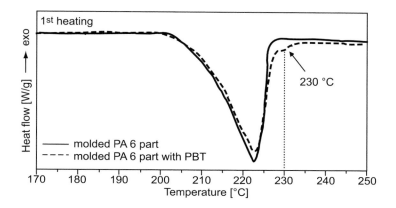

Fig. 1.85   Melting point curves of two molded PA 6 parts, with and without PBT contamination

1st Heating scan, specimen mass approx. 10 mg, heating rate 10 °C/min, purge gas: nitrogen

Because the PA 6 and PBT parts were injection molded on the same machine, it seemed reasonable to conclude that the polyamide might have been contaminated by residual PBT from the dryer or the metering device. The effect in the curve is comparatively small and superposed by the PA melting point curve. For this reason, an additional cooling curve was recorded for the contaminated molded part, Fig. 1.86.

This clearly reveals two crystallization peaks that can be resolved better from each other than the corresponding melting peaks obtained by heating. The crystallization peak at T_{pc} = 190 °C can be assigned to the PA 6 by reference to that of the uncontaminated part. The crystallization peak occurring at the higher temperature of T_{pc} = 194 °C is caused by the PBT contaminant.

The premature crystallization brought about by the PBT contaminant may have caused the gate to freeze as the material was being injected. Consequently, because the holding pressure was no longer effective, voids probably formed in the molded part.

Semicrystalline contamination often shows up much more clearly in the cooling curve than in the heating curve.

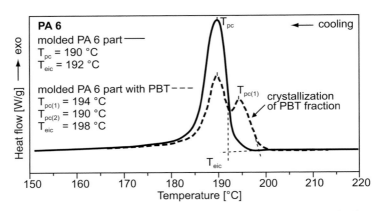

Fig. 1.86   Crystallization of molded PA 6 parts, with and without PBT contamination

T_{eic} = Crystallization extrapolated onset temperature,

T_{pc} = crystallization peak temperature,

Specimen mass approx. 10 mg, cooling rate 10 °C/min, purge gas: nitrogen

1.2.3.9   Curing of Thermosets

When DSC is used on thermosets (reactive resins), essentially two effects are measured:
- the endothermic glass transition, and,
- the exothermic crosslinking enthalpy.

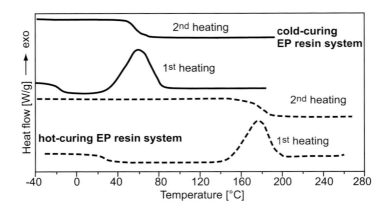

Fig. 1.87   1st and 2nd heating scan of cold-curing and hot-curing EP resin systems

Specimen mass approx. 15 mg, heating rate 10 °C/min, purge gas: nitrogen

. Whether crosslinking occurs at ambient temperature or at elevated temperatures on the type of resin system. Such systems are called cold-curing and hot-curing systems respectively, Fig. 1.87. The glass transition temperature of heat-curing systems is correspondingly higher.

Unlike curing in real life, which is frequently performed in an isothermal atmosphere, curing in this example occurred during heating. This is why the curing peaks are displaced toward higher temperatures. Cold-curing systems exhibit a glass transition at about -20 °C which is more or less that of the uncrosslinked resin. Exothermic curing starts at around 40 °C. The 2nd heating scan (and perhaps, for confirmation, a 3rd one) produces no further reaction peaks. Consequently, the glass transition temperature may be used to determine the maximum degree of curing attainable (often referred to as "full cure") under practical conditions. A prerequisite for this is that the end temperature of the 1st heating scan is not so high that degradation and thus a lowering of T_g occurs. Nor must it be so low (i.e., the whole reaction peak is not obtained in the 1st heating scan) that full curing does not take place. Hot-curing systems crosslink at elevated temperatures. They therefore have a higher glass transition temperature and higher service temperatures.

The start of curing reveals a dependence not only on the temperature but also on time. For this reason, when DSC is used to measure the curing of reactive resins, the course of the reaction is also found to depend on the heating rate. Low heating rates, as opposed to high ones, trigger the reaction at lower temperatures. However, if curing is performed isothermally, the higher the holding temperature, the earlier curing will start.

The magnitude of the reaction enthalpy is also influenced by the measuring parameters (heating rate, isothermal holding temperature). Changes in reaction behavior are most likely to occur at very low heating rates or at isothermal holding temperatures.

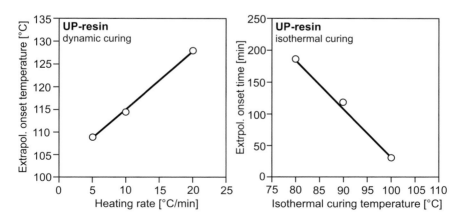

Fig. 1.88   Curing of UP-resin; reaction begin as a function of heating rate
                    or isothermal curing temperature

Specimen mass approx. 6 mg, purge gas: nitrogen, pressure pan

The **degree of curing** c_{DSC} of a molded material is obtained by dividing the released residual reaction enthalpy ΔH_r by the total reaction enthalpy $\Delta H_{tot.}$ of a freshly prepared reactive resin compound of the same composition. The degree of curing of a resin system critically affects a number of the molded part's properties, such as rigidity and chemical resistance.

$$c_{DSC} = \left(1 - \frac{\Delta H_r}{\Delta H_{tot}}\right) \cdot 100 \quad [\%]$$

c_{DSC}	–	*Degree of curing from DSC measurement*
ΔH_{tot}	–	*Total reaction enthalpy*
ΔH_r	–	*Released residual reaction enthalpy*

In the case of filled or fiber-reinforced materials, the fraction of reactive polymer present must be determined (by ashing or thermogravimetry) so as to permit the degree of curing to be quantified.

> **Degree of curing influences the properties of a molded part,
> for example, rigidity and chemical resistance.**

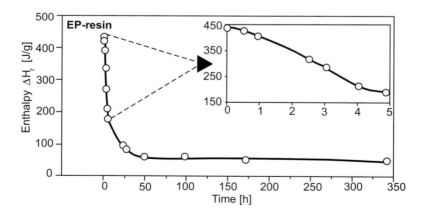

Fig. 1.89   Change in residual enthalpy of cold-curing EP-resin after different periods of storage

Elapsed time and the temperature during curing contribute immensely to the degree of curing. Figure 1.89 shows for a cold-curing EP resin stored for different lengths of time at ambient temperatures after preparation the decrease in that reaction enthalpy that is amenable to determination by DSC. The greatest enthalpy change occurs within the first 24 hours. The reaction slows down afterwards but, even after 2 weeks, DSC still registers a residual reaction. Faster, full curing of this resin system is thus feasible only by appropriate **postcuring** at elevated temperatures.

DSC frequently fails to detect significant residual reaction in extensively cured reactive resins as by then only a few bonds are still being formed, which means that little heat is being evolved. Nevertheless, because the formation of these "last" bonds greatly increases the crosslinking density, the glass transition temperature continues to rise. At this advanced stage, it is therefore advisable to use T_g to characterize the degree of curing. Figure 1.90 shows the relationship between the T_g and the degree of curing, determined via the residual enthalpy, of a vinyl ester resin postcured under various conditions (different lengths of time at different temperatures). T_g increases disproportionately in the region of fuller curing.

> **A rise in T_g indicates further curing, even if residual enthalpy**
> **can no longer be detected.**

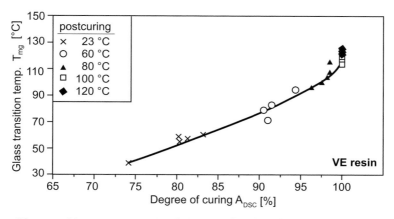

Fig. 1.90   Glass transition temperature T_{mg} (DSC) as a function of the degree of curing c_{DSC}
               from residual enthalpy measurements for a VE resin cured in various conditions [35]

Aside from permitting the degree of curing to be assessed, DSC can identify mixing errors, especially in the case of epoxy resins. When a certain set temperature is not reached during or after the 2nd heating scan at the latest, a false mixing ratio is the likely cause.

Volatile reaction partners, like styrene in UP resin, make it difficult to establish the reaction enthalpies. Styrene may escape during determination of the total reaction enthalpy, for example, of a freshly prepared UP resin mixture, and will thus not participate further in curing during the heating process. Using pressure pans will stop the styrene escaping (see Section 3.2.2.1).

Figure 1.91 illustrates this for the crosslinking reaction of an SMC paste in both a normal and a pressure pan. A total reaction enthalpy of 177 J/g is measured in the pressure pan, more than twice the value of that in the normal pan (87 J/g). This source of error must be borne in mind when total reaction enthalpies are being determined. But, again, it must also be remembered that the increased pressure in the pressure pan influences the curing reaction and therefore also the enthalpy.

> **Hermetically sealable pans must be used for volatile substances.**

Another point to be taken into account when thermosets are being used is the **storage stability** of the reactive mixtures. Warm-curing and hot-curing ready-to-process resin systems can be deep frozen for a certain length of time. Even so, slight crosslinking may still occur during such storage. This is illustrated in Fig. 1.91, which shows the DSC curves of two SMC paste specimens.

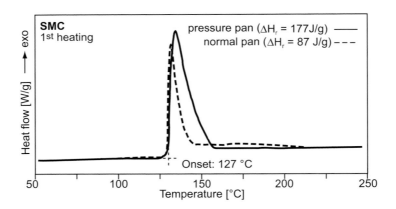

Fig. 1.91   Crosslinking reaction of SMC paste in normal and pressure pans

1ˢᵗ Heating scan, specimen mass approx. 20 mg, heating rate 10°C/min,
purge gas: nitrogen

The freshly prepared SMC paste has a reaction enthalpy of 177 J/g. This is 26% higher than that of the stored batch (2 years at -20 °C), whose lid was removed periodically to allow sampling. In addition, the heating curve shows an endothermic effect between 70 and 100 °C. This is due to the evaporation of water that condensed on the surface of the cold specimen during opening.

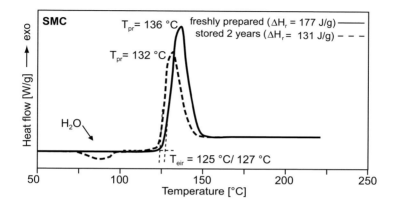

Fig. 1.92   Crosslinking reaction of SMC paste, freshly prepared and stored for 2 years at -20 °C

1st Heating scan in the pressure pan, specimen mass approx. 20 mg,
heating rate 10 °C/min, purge gas: nitrogen

In view of the problems described in the preceding, it is usual to mix resin-hardener systems just prior to processing. This approach must also be adopted for DSC determinations of total reaction enthalpy. The effect of storage at different temperatures (refrigerator, ambient, elevated ambient, for example, in summer) on the total reaction enthalpy is shown in Fig. 1.93.

The prepared resin/hardener system was stored at different temperatures for different periods. Even storage in the refrigerator leads to postcuring and invalidates the results for the desired initial state. This is particularly true of cold-curing resins, as these start to react at relatively low temperatures.

Measure freshly prepared resin/hardener system immediately.
Even protracted storage at low temperatures does not prevent the reaction
from starting (applies especially to cold-curing resin systems).

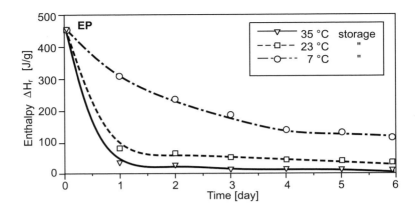

Fig. 1.93   Influence of storage temperature and time on the reaction enthalpy of a
               cold-curing EP resin

1.2.3.10 Results of Round-Robin Trials

Round-robin trials afford a means of assessing the comparability of experimental results and the measuring reliability of a laboratory. In a round-robin, defined specimens are tested and evaluated in different laboratories at the same time and according to agreed procedures.

The degree of agreement between independent test results obtained by the same analyst on the same sample by the same method and using the same equipment in the same laboratory within a short time period is defined as the **repeatability, r**.

When identical samples are analyzed by the same method under comparable conditions, and different analysts use different equipment, usually in different laboratories, to evaluate the results, this is referred to as **comparability, R**.

Determination of the Glass Transition Temperature in Accordance with DIN 53765

Amorphous and semicrystalline specimens were analyzed, namely PMMA, PC, and PSU. The results indicated that the limits of repeatability and comparability are very narrow: the repeatability r (same laboratory) ranges from 1.0 °C to 2.0 °C, and the comparability R (different laboratory) lies between 3.0 °C and 5.0 °C [49]. This means for practical purposes that the glass-transition temperatures in the same place differ by at most 1–2 °C and in different places, by at most 3–5 °C, Table 1.11.

Additional evaluation of the data in accordance with a method proposed in the literature, namely averaging the glass transition temperatures from the second heating scan and from cooling, failed to significantly improve the repeatability or the comparability [49].

Specimen	Result	Mean T_{mg} [°C]	Repeatability, r [°C]*	Comparability, R [°C]*
PMMA	1st Heating scan	103.8	1.99	5.26
	Cooling	96.5	2.10	5.04
	2nd Heating scan	103.0	1.32	3.92
PC	1st Heating scan	148.3	1.51	2.58
	Cooling	142.1	1.99	4.06
	2nd Heating scan	146.9	1.54	3.22
PSU	1st Heating scan	187.4	1.90	3.22
	Cooling	181.5	1.68	4.37
	2nd Heating scan	187.1	1.12	3.22

*Temperatures quoted to one or two decimal places have been taken from the literature. The round-robin results also showed that it is sufficient to quote the temperature for polymers to the nearest °C.

Table 1.11   Round-robin determination of the glass transition temperature T_{mg} of the repeatability r (same laboratory) and the comparability R (different laboratory) for various thermoplastics [49]

Change in Specific Heat Capacity (Δc_p)

The step height of T_g measurements can be used to determine the change in specific heat capacity (Δc_p) as well as the actual glass transition temperature. A high degree of similarity was observed for the repeatability r of the 2nd heating scans on the various specimens. The comparability R was not so sharply defined. Results from the 2nd heating scan are presented in Table 1.12.

Specimen	Specific heat capacity Δc_p [J/(K g)]	Repeatability, r [J/(K g)]	Comparability, R [J/(K g)]
PMMA	0.309	0.048	0.126
PC	0.235	0.039	0.081
PSU	0.225	0.042	0.078

Table 1.12   Round-robin determination of the specific heat capacity Δc_p of the repeatability r (same laboratory) and the comparability R (different laboratory) for different thermoplastics [49]

Comparison of the values from the 2nd heating scans reveals that, in this round-robin trial, power-compensation DSC instruments tended to return lower mean readings than heat-flux instruments. It is not possible to establish on the basis of this round-robin trial if the readings do in fact depend on the underlying measuring method, and further investigation is needed [49].

Crystallinity of Semicrystalline Thermoplastics

Round-robin trials on polyethylenes have shown that, at starting temperatures of 25 °C, the start-up deflection of the curve does not constitute a definite baseline and thus definite degrees of crystallization cannot be determined. When measurements are conducted under comparability and repeatability conditions (e.g., for quality control), it is advisable to set the starting and onset temperatures as low as possible in advance [49, 50].

1.2.3.11   Compiling TTT Diagrams

As already shown in Section 1.2.3.9, resin curing can be studied with the aid of DSC. The isothermal holding temperature is used for establishing the kinetics of the reaction, that is, the degree of curing as a function of time and temperature, and a glass transition temperature is assigned to each degree of cure.

A time/temperature/transition diagram (TTT), based on DSC characteristic values inter alia, provides a schematic overview of isothermal curing (Fig. 1.94). A TTT diagram is a plot of temperature, which may be regarded as the curing temperature, against time. Any horizontal line from left to right in the diagram corresponds to isothermal curing at the equivalent temperature.

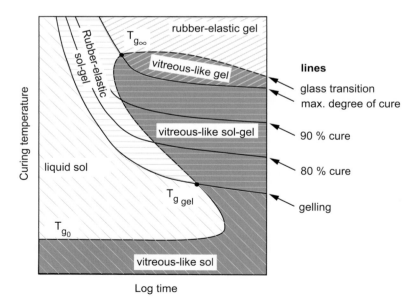

Fig. 1.94   Schematic illustration of a TTT diagram [51, 52]

As the resin cures, it passes through the three morphological states: namely liquid, gel-like (rubber-elastic) and solid (vitreous).

Whereas gelling represents the limit of processability, the glass transition temperature connotes the incipient application range of the finished part.

The transition from liquid to a gel-like state is determined experimentally from the rheological response of the resin, for example, from isothermal measurements made with a rotary viscometer. DSC can additionally assist in the determination of the degree of curing at every point in time at any given temperature. In practice, numerical methods are used to compute the reaction kinetics from at least three isothermal measurements (Fig. 1.95); this process yields what are known as "lines of degree of cure."

Because there are laws governing the relationship between degree of cure and glass-transition temperature, it is possible to construct the line corresponding to the transition to the gel-like state. In addition, DSC can be used to determine the glass transition temperatures T_{g0} of the fully uncured resin (1st heating scan) and the $T_{g\infty}$ of the fully cured resin (2nd or 3rd heating scan).

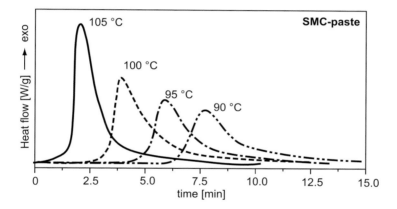

Fig. 1.95   Isothermal reaction curves for an SMC paste measured at different curing temperatures for the purpose of constructing a TTT diagram

Specimen mass approx. 10 mg, purge gas: nitrogen

The appearance of a TTT diagram may vary considerably according to the type of resin concerned. With epoxy resins, for example, the polyaddition mechanism by which the comparatively short oligomers undergo stepwise linkage causes gelling to occur at high degrees of cure (often in excess of 50%). In contrast, UP resins, which cure by means of a chain reaction, mostly react faster and gel very quickly (but the percentage conversion is lower) because the starting molecules are long and contain highly functional groups. The shape of the TTT diagram is also largely determined by the functionality within a particular type of resin, that is, by the number of reactive groups per resin molecule.

When TTT diagrams are being examined, it should be noted that they are ideal representations. In other words, they do not allow for exothermic reactions of the kind that occur in practice or for temperature programs.

There are several things to bear in mind when attempting to determine reliable, reproducible values for a TTT diagram. These stem from the often numerous and sometimes readily volatile additives present and the risk of reaction between the starting materials and atmospheric humidity:

- Where possible, one technique should be used for determining the various characteristic values (DMA would often be a candidate for determining the glass-transition temperature of almost fullycured substances, but is difficult to use for freshly prepared resin systems; accordingly, DSC is usually employed for determining the glass-transition temperature of all states),
- Use pressure pans so that evaporation of readily volatile additives may be minimized,
- Choose a suitable heating rate for the instrument in establishing the isothermal curing temperature (40–60 °C/min); if it is too high, it will cause the furnace to overshoot and thus create undefined reaction conditions; if it is too low, it will lead to premature curing of the resin during the dynamic heating phase,
- Correct, reproducible conditions during preparation of the fresh resin system (storage in refrigerator, choice of suitable batch size, ensuring homogeneous mixing, especially of highly filled systems).

1.2.3.12   Examples of Temperature-Modulated DSC (TMDSC)

Temperature-modulated DSC is particularly suitable when overlapping effects occur, glass transitions are hard to see, and standard DSC yields a signal that is difficult to interpret, if at all. Where irreversible and reversible processes occur and overlap, TMDSC can be used to separate them. Frequently, reversible, repeatable effects (glass transitions, melting and crystallization peaks) are superimposed with irreversible, nonrepeatable effects (postcuring peaks, molecular orientation). While such irreversible effects can be eliminated with conventional methods by a combination of 1st heating scan, cooling and 2nd heating scan, this can alter the specimen material; for example, it might increase the glass transition temperature of crosslinking polymers. An additional advantage of using TMDSC is that the signals from the separated transitions are much more pronounced than in standard DSC.

**TMDSC separates superposed effects
(thermal and mechanical history of a material property) in the 1st heating scan.**

Figure 1.96 shows the TMDSC heating curve for ABS pellets. The solid curve represents the cumulative curve of the overall change in heat flux; that is, it is the curve produced by a standard DSC experiment.

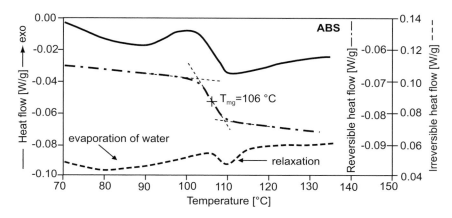

Fig. 1.96   Determination of glass transition temperature T_{mg} of ABS by means of TMDSC; overlapping of curves is due to evaporation of water and relaxation

Specimen mass approx. 10 mg, heating rate 2 °C/min, amplitude +/- 0.5 °C, period 60 s, purge gas: nitrogen

The unsteady baseline before T_{mg} (106 °C) and the transition after T_{mg} make it difficult to evaluate the glass transition temperature. However, the latter can be clearly evaluated from the reversible heat flux curve. The irreversible heat flux curve reveals two endothermic effects, namely the evaporation of water between 60 °C and 100 °C and relaxation of orientation at 105 °C. These two effects occur only in the 1st heating curve and are not re-peatable.

In a mixture of PEKEKK and PES (60 : 40 parts by weight), the overall heat flux signal is dominated by exothermic postcrystallization of the semicrystalline PEKEKK, Fig. 1.97. The reversible heat flux reveals the glass transition temperatures of the semicrystalline PEKEKK at 161 °C and of the purely amorphous PES at 221 °C. The irreversible heat flux emphasizes the exothermic postcrystallization of the PEKEKK.

Figure 1.98 illustrates a postcuring reaction of an incompletely cured molded EP resin part superposed on the glass transition. The overall heat flux reveals only the exothermic curing reaction. The reversible heat flux curve makes it possible to evaluate the glass transition, which is in the same temperature range, with a midpoint temperature of T_{mg} = 73 °C. The ir-reversible heat flux reveals only the postcuring reaction.

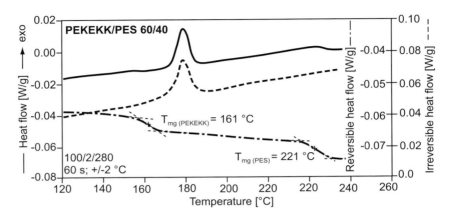

Fig. 1.97   Glass transitions and postcrystallization of a PEKEKK/PES mixture (60 : 40) revealed by TMDSC

Specimen mass approx. 7 mg, heating rate 2 °C/min, amplitude +/- 0.5 °C, period 60 s, purge gas: nitrogen

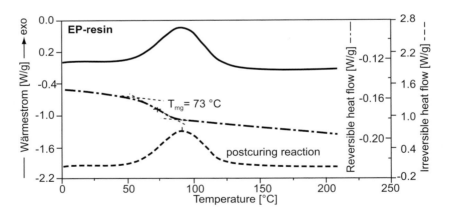

Fig. 1.98   Postcuring reaction and glass transition of an EP resin revealed by TMDSC

Specimen mass approx. 25 mg, heating rate 2 °C/min, amplitude +/-0.5 °C, period 60 s, purge gas: nitrogen

1.3 References

[1] N.N. ISO 11357-1 (1997)
Plastics – Differential Scanning Calorimetry (DSC) –
Part 1: General Principles

[2] N.N. DIN 51 005 (1999)
Thermal Analysis (TA) – Terms/ Note: Intended as
Replacement for DIN 51 005 (1993)

[3] N.N. ASTM D 3417 (1999)
Standard Test Method for Enthalpies of Fusion and
Crystallization of Polymers by Differential Scanning
Calorimetry

[4] N.N. DIN 53 765 (1994)
Testing of Plastics and Elastomers – Thermal Analysis;
DSC-Method/ Note: EQV ISO 11357-5 (1999)

[5] Ehrenstein, G.W. Polymeric Materials
Carl Hanser Publishers, Munich 2001

[6] N.N. ISO 11357-2 (1999)
Plastics – Differential Scanning Calorimetry (DSC) –
Part 2: Determination of Glass Transition Temperature

[7] N.N. ASTM E1356 (2003)
Standard Test Method for Assignment of the Glass
Transition Temperatures by Differential Scanning
Calorimetry

[8] N.N. ASTM D 3418 (1999)
Standard Test Method for Transition Temperatures by
Differential Scanning Calorimetry Analysis

[9] N.N. ASTM E 794 (2001)
Standard Test Method for Melting Temperatures and
Crystallization Temperatures by Thermal Analysis

[10] Rieger, J.   Bestimmung der Glastemperatur mittels DSC
                                          Kunststoffe 85 (1995) 4, pp. 528–530

[11] Könnecke, K.   Personal Information, February 1998

[12] Henning, I.   Personal Information, February 1998

[13] Wingfield, M.   Personal Information, February 1998

[14] Wunderlich, B.   Thermal Analysis
                                          Academic Press, San Diego, 1990

[15] Illers, K.-H.   Die Ermittlung des Schmelzpunktes von kristallinen
                                          Polymeren mittels Wärmeflußkalorimetrie (DSC)
                                          European Polymer Journ. (1974) 10, pp. 911–916

[16] Varga, J.,   Isothermal Crystallization of the β-Modification of
      Garzo, G.   Polypropylene
                                          Acta Chimica Hungarica 128 (1991), pp. 303–317

[17] Wunderlich, B.   Advanced Thermal Analysis Laboratory (ATHAS),
                                          Bept. Chem. University Tennessee, Knoxville,
                                          TN 37996-1600, USA

[18] van Krevelen, D.W.   Properties of Polymers
                                          Elsevier, Amsterdam, 1997

[19] Rieger, J.   Personal Information, 1992

[20] N.N.   ASTM E 793 (2001)
                                          Standard Test Method for Enthalpies of Fusion and
                                          Crystallization by Differential Scanning Calorimetry

[21] N.N.   ISO 11357-3 (1999)
                                          Plastics – Differential Scanning Calorimetry (DSC) –
                                          Part 3: Determination of Temperature and Enthalpy of
                                          Melting and Crystallization / Note: To be amended by
                                          ISO 11357-3 DAM 1(2004)

[22] Wunderlich, B.   Makromolecular Physics – Vol. II
                                          Academic Press New York, San Fransisco, London 1976

[23]   N.N.   ISO 11357-5 (1999)
                                     Plastics – Differential Scanning Calorimetry (DSC) –
                                     Part 5: Determination of Characteristic Reaction Curve
                                     Temperatures and Times, Enthalpie of Reaction and
                                     Degree of Conversion / Note: EQV DIN 53765 (1994)

[24]   N.N.   ASTM E 1269 (2001)
                                     Standard Test Method of Determining Specific Heat
                                     Capacity by Differential Scanning Calorimetry

[25]   N.N.   ISO/DIS 11357-4 (2003)
                                     Plastics – Differential Scanning Calorimetry (DSC) –
                                     Part 4: Determination of Specific Heat Capacity

[26]   Widmann, G.,   Thermoanalyse
       Riesen, R.   Hüthig Buch Verlag GmbH, Heidelberg 1990

[27]   Höhne, G.W.H., u.a.   Temperature Calibration of Differential Calorimeters
                                     PTB-Information 100, January 1990

[28]   N.N.   ASTM E 967 (2003)
                                     Standard Practice for Temperature Calibration of
                                     Differential Scanning Calorimeters and Differential
                                     Thermal Analyzers

[29]   N.N.   ASTM E 2069 (2000)
                                     Standard Test Method for Temperature Calibration on
                                     Cooling of Differential Scanning Calorimeters

[30]   Sarge, S.M., u.a.   Thermal Calibration of Differential Calorimeters
                                     PTB-Information 103, June 1993

[31]   N.N.   ASTM E 968 (2002)
                                     Standard Practice for Heat Flow Calibration of
                                     Differential Scanning Calorimeters

[32]   N.N.   DIN 51 007 (1994)
                                     Thermal Analysis – Differential Thermal Analysis;
                                     Principles

[33]   N.N.   TA Instruments GmbH
                                     Operator's Guide DSC 2920
                                     TA Instruments, Alzenau 1997

[34] Turi, E.A.   Thermal Characterisation of Polymeric Materials,
                                      Second Edition
                                      Academic Press, San Diego 1997

[35] Ehrenstein, G.W.,   Duroplaste
      Bittmann, E.   Aushärtung, Prüfung, Eigenschaften
                                      Carl Hanser Publishers, Munich 1997

[36] Oberbach, K.,   Saechtling: Kunststofftaschenbuch, 29th Edition
      Baur, E.,   Carl Hanser Publishers, Munich 2004
      Brinkmann, S.,
      Schmachtenberg, E.

[37] Widmann, G.,   Collected Applications of Thermal Analysis
      Jandali, M.   Thermoplaste, Mettler-Toledo GmbH (1997), p. 19

[38] Domininghaus, H.   Die Kunststoffe und ihre Eigenschaften, 5th Edition
                                      Springer Verlag, Berlin 1999

[39] Ehrenstein, G.W.   Schadensanalyse
                                      Carl Hanser Publishers, Munich 1992

[40] N.N.   DIN 50 035 (1989)
                                      Terms and Definitions used on Ageing of Materials

[41] Pongratz, S.   Thermoanalytical Methods for Characterizing Aging
                                      Proceeding: Thermal Methods in Plastics Testing
                                      Erlangen 1997

[42] Mathot, V.B.F.   Calorimetry and Thermal Analysis of Polymers
                                      Carl Hanser Publishers, Munich 1994

[43] Thomas, L. C.   Use of Multiple Heating Rate DSC and Modulated
                                      Temperature DSC to Detect and Analyze
                                      Temperature-Time-Dependent Transition in Materials

[44] Gill, P.S.,   Modulated Differential Scanning Calorimetry
      Sauerbrunn, S.R,   Journal of Thermal Analy. 40 (1993) pp. 931–939
      Reading, M.

[45] Wunderlich, B.,   Mathematical Description of Differential Scanning
      Jin, Y.,   Calorimetry Based on Periodic Temperature Modulation
      Boller, A.   Thermochim. Acta 238 (1994) pp. 277–293

[46]   Schawe, J.E.K   Modulated Temperature DSC Measurements -
                                 the Influence of Experimental Conditions
                                 Universität Ulm, Sektion für Kalorimetrie

[47]   Szameitat, M.   Reibung and Verschleiß von Mischungen aus
                                 hochtemperaturbeständigen thermoplastischen
                                 Kunststoffen,
                                 Dissertation Universität Erlangen – Nürnberg 1998

[48]   Möhler, H.   Die Thermische Analyse in der Kunststoffprüfung
                                 Kunststoffe 84 (1994) 6, pp. 736–743

[49]   Affolter, S.,   Interlaboratory Tests on Polymers: Thermal Analysis
       Schmid, M.   Int. J. Polym. Anal. Charact. 6 (2000), p. 35

[50]   Affolter, S.,   Interlaboratory Tests on Polymers by Differential
       Ritter, A.,   Scanning Calorimetry (DSC): Determination of Glass
       Schmid, M.   Transition Temperature (T_g)
                                 Macromol. Mater. Eng. 286 (2001), p. 605

[51]   Ehrenstein, G.W.,   Curing of Reaction Resins: The Time-Temperature-
       Bittmann, E.,   Transition-Diagram
       Wacker, M.   http://www.lkt.uni-erlangen.de, Feb. 2003

[52]   Hadiprajitno, S.,   Developing Time-Temperature-Transformation
       Hernandez, J.P.,   Diagrams for Unsaturated Polyesters Using DSC Data
       Osswald, T.   SPE Proceedings ANTEC, Nashville/USA (2003),
                                 pp. 818–822

[53]   Bryant, E.   Crosslink Density and the Glass Transition in
                                 Thermosetting Polymers
                                 Term Paper, CME 509,
                                 www.udri.udayton.edu/rpdl/Paper_Crosdens.htm

Other references:

[54]   N.N.   ISO 11357-7 (2002)
                                 Plastics – Differential Scanning Calorimetry (DSC) –
                                 Part 7: Determination of Crystallization Kinetics

[55]   N.N.   ISO/DIS 11357-8 (2001)
                                 Plastics – Differential Scanning Calorimetry (DSC) –
                                 Part 8: Determination of

[56] N.N.   ASTM E 2070 (2000)
                                    Standard Test Method for Kinetic Parameters by
                                    Differential Scanning Calorimetry Using Isothermal
                                    Methods

[57] N.N.   ASTM D 4591 (2001)
                                    Standard Test Method of Determining Temperatures and
                                    Heats of Transitions of Fluoropolymers by Differential
                                    Scanning Calorimetry

[58] N.N.   prEN 6041 (1995)
                                    Aerospace Series – Non-metallic Materials –
                                    Test Method; Analysis of Non-metallic
                                    (uncured) by Differential Scanning Calorimetry (DSC)

[59] N.N.   prEN 6064 (1995)
                                    Aerospace Series – Non-metallic Materials – Analysis of
                                    Non-metallic Materials (cured) for the Determination of
                                    the Extend of Cure by Differential Scanning Calorimetry
                                    (DSC)

[60] Hemminger, W.F.,   Methoden der Thermischen Analyse
      Cammenga, H.K.   Springer-Verlag Berlin, Heidelberg 1989

[61] Höhne, G.W.H.,   Differential Scanning Calorimeter
      Hemminger, W.,   An Introduction for Practitioners
      Flammersheim, H.-J.   Springer Verlag, Heidelberg 1997

[62] Merzlyakov, M.,   Complex Heat Capacity Measurements by TMDSC
      Schick, C.   Influence of non-linear Thermal Response
                                    Thermochimica Acta 330 (1999), pp. 55–64

[63] Hensel, A.,   Temperature Calibration of Temperature-Modulated
      Schick, C.   Differential Scanning Calorimeters
                                    Thermochimica Acta 304/305 (1997), pp. 229–237

[64] Schawe, J. E. K.   Modulated Temperature DSC Measurements,
                                    The Influence of the Experimental Conditions
                                    Thermochimica Acta 271 (1996), pp. 127–140

[65] Boller, A.,   Modulated Differential Scanning Calorimetry in the
      Schick, C.,   Glass Transition Region
      Wunderlich, B.   Thermochimica Acta 266 (1995), pp. 97–111

2   Oxidative Induction Time/Temperature (OIT)

2.1   Principles of OIT

2.1.1   Introduction

Plastics age throughout their service lives. OIT is a method that uses differential scanning calorimetry (DSC) to make comparative inferences about the stability of materials to thermooxidative attack.

Aging (see Section 1.2.3.6) can be hindered or retarded by the addition of stabilizers. These are chemical substances that modifiy the degradation mechanism and stop or at least delay incipient chain reactions. The most important stabilizers offer protection against oxidation, heat, light, and radiation.

**Important stabilizers:
Antioxidants, processing stabilizers, photoprotective agents**

Most plastics need to be stabilized against thermooxidative degradation. Phosphites and phosphonites generally serve as stabilizers when processing places high, short-term thermal stress on melts in the presence of oxygen. Because processing temperatures can be 200 °C or higher than the service temperature, short-term stabilizers have to meet different demands than long-term stabilizers, which are used for plastics in the solid state and which are intended to counteract degradation over long-term use at service temperatures. Phenols often serve as short-term stabilizers [1].

Photoprotective agents such as UV absorbers and sterically hindered amines hindered amine light stabilizer (HALS) are also used. These are usually effective in concentrations as low as fractions of 1%. Generally, a system of different stabilizers is used so as to obtain a synergistic effect. However, this is not absolutely necessary [2].

The effectiveness of stabilizers is assessed by aging stabilized plastics in drying cupboards and then testing various characteristics. OIT is often employed in DSC because specimen preparation is simple and the method is fast. However, it is not advisable to extrapolate the conclusions to predict long-term stability or to use the method for a comparative study of polymers stabilized with different chemicals.

**Rapid DSC method for determining stabilizer effectiveness
(assuming stabilizers are the same)**

OIT results cannot be used directly to determine suitability for service or the effectiveness of stabilizers in practice. Stabilized polypropylene (PP) serves as an example of the problems that may arise:

- Phosphites and phosphonites give rise to long OIT times at high test temperatures but are much less effective at actual service temperatures (which are far lower).
- Low molecular weight phenolic compounds are volatile under OIT test conditions but they can contribute to long-term heat stability at elevated service or test temperatures.
- In the case of high molecular weight antioxidants (mostly sterically hindered phenols), the OIT correlates well with the concentration of the antioxidant. However, OIT values cannot be extrapolated to long-term behavior.
- Photostabilizers (HALS) based on sterically hindered amines are not active radical scavengers at test temperatures of around 200 °C. However, in oven-aging trials at a typical temperature of 100 °C, they make a significant contribution to long-term stability [3, 4].

OIT times cannot be used to make direct estimates of the long-term effectiveness of stabilizers in practice.

Other aspects of OIT to be noted are the following:

OIT measurements are taken when the polymer is molten, that is, at temperatures roughly on a par with melting/processing temperatures, but far above normal service temperatures. In semicrystalline thermoplastics, the stabilizers accumulate in the amorphous domains. When the crystallites melt, the stabilizers suddenly become much more mobile and soluble and disperse throughout the melt [5–7].

When a phenol/phosphite stabilizer mixture is aged in an oven at 149 °C, the phosphite is converted into phosphate within several hours. Consequently, only phenols are responsible for the aging time. By contrast, in a phenol/thiosynergist mixture, it is the thiosynergist content that determines aging. The OIT values, however, do not reveal this [7].

In oven aging, specimen thickness and external stresses together with ambient air and media flow all exert an influence.

2.1.2 Measuring Principle

Because the reaction between polymers and oxygen is exothermic, DSC is an ideal method for studying this process. The experiment consists of exposing the specimen to a gaseous oxidant (oxygen or air) at elevated temperatures. Two variants are commonly employed:

- The dynamic method, which yields a characteristic oxidation induction temperature "OIT temperature",
- The static method, which yields a characteristic oxidation induction time "OIT time".

In the **dynamic method**, the specimen is heated in an oxygen or air atmosphere from the very start. A shift from the baseline that occurs at a characteristic temperature is a clear indication that the oxidation reaction is exothermic.

This method has the advantage of being fast but is less sensitive because the temperature continues to rise. It cannot therefore be used to distinguish slight differences in stabilizer effectiveness. The dynamic method can be used for almost all plastics.

Dynamic OIT
Less sensitive, but faster – Suitable for almost all plastics

In the **static method**, which is standardized in DS 2131.2 [8], ASTM D 3895 [9] and ISO 11357-6 [10], the specimen is heated in an inert atmosphere to a specified temperature above the melting point. This temperature is then kept static. Once equilibrium has been attained, the atmosphere is switched from inert to oxidative, usually from nitrogen to oxygen or air.

The exothermic oxidative reaction usually occurs after a certain time interval. This method has the advantage of revealing fine gradations in the degree of stabilization.

To establish an appropriate test temperature, preliminary dynamic OIT trials need to be carried out; these take a long time. The static method is preferred for plastics exposed to long-term heat, for example polyolefins in pipelines and in equipment making.

Static OIT
More sensitive, but more time consuming
Normally used for polyolefins (because it is standardized)

2.1.3 Procedure and Influential Factors

The stages in an OIT measurement are:

- Prepare the specimen.
- Weigh out the specimen into an **open** pan.
- Place the specimen and reference pans in the test chamber.
- Set the gas flow.
- Select the appropriate heating program.
- Switch to an oxidative atmosphere (in the static method).

Note: *OIT measurements are occasionally performed at elevated pressure in a pressure-DSC instrument to simulate mechanical processing. This type of measurement is not discussed here.*

Instrumental and specimen-specific factors are:

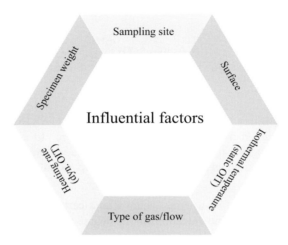

Actual experimental curves are presented in Section 2.2.2 to explain in detail the influential factors and sources of error during an experiment.

2.1.4 Evaluation

2.1.4.1   Dynamic Method

In dynamic OIT, the specimen is heated continuously and the specimen chamber is purged from the start with oxidative gas.

Figure 2.1 is a typical experimental curve for a semicrystalline thermoplastic. The first transition is endothermic melting. It is followed at higher temperatures by an exothermic baseline shift (peak) signaling the start of oxidation.

The OIT temperature is usually taken to be the extrapolated onset temperature T_{eio}. This is obtained by extending the baseline beyond the start of the exothermic peak to intercept the extrapolated steepest linear slope of the peak. If the exothermic peak has a leading edge or the onset temperatures are difficult to reproduce, it is preferable to use a temperature T_x^{dy}, which is the change in heat flux offset from the baseline by a value of x; frequently a value of 0.2 W/g is used.

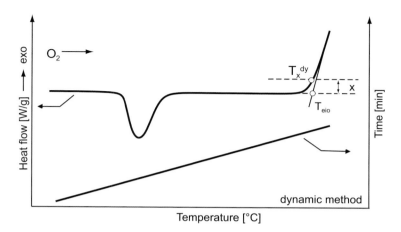

Fig. 2.1   Evaluation of a dynamic OIT curve in accordance with ISO 11357-6 [10].
                Labels according to ISO 11357-1 [11]

> T_{eio} = *Extrapolated oxidative onset temperature, intersection of the tangents before*
> *and during oxidation*
>
> T_x^{dy} = *Temperature after exothermic offset of the heat flux from the baseline by x*

2.1.4.2   Static Method

A specimen is heated above its melting point and held of this temperature, Fig. 2.2. A disturbance in the heat flux cure occurs due to the reduction in heating rate from 10 °C/min to isothermal conditions. After a delay of 1–3 minutes while steady-state is achieved, the atmosphere is changed from inert to oxidizing.

As in the dynamic method, evaluation of the exothermic shift from the baseline takes the form of recording the OIT times, the extrapolated onset time t_{eio} and a constant offset from the baseline t_x^{stat}. Again, the time t_x^{stat} is preferred where leading edges occur or the onset times are difficult to determine.

To enable comparisons, the change in heat flux x must be kept constant while T_x^{dy} or t_x^{stat} is being evaluated. No reliable transition temperatures or enthalpies can be recorded for measurement in an open pan because good thermal contact between the specimen and the pan cannot be guaranteed.

> **Measurements in an open pan enables exact details of transition temperatures or enthalpies may be made.**

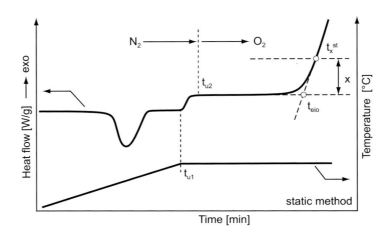

Fig. 2.2   Evaluation of a static OIT curve in accordance with DS 2131.2 [8], ASTM D 3895 [9]
                and ISO 11357-6 [10]. Labels according to ISO 11357-1 [11].

    t_{U1} = *Switch to isothermal temperature*

    t_{U2} = *Switch from inert to oxidative gas atmosphere*

    t_{eio} = *Extrapolated oxidative onset time; intersection of tangent before and during*
                oxidation

    t_x^{stat} = *Temperature after exothermic offset of the heat flux from the baseline by x*

2.1.4.3   Test Report

The test report contains the same details as for DSC (see Section 1.1.4.6), taking the various
OIT methods into account.

2.1.5 Overview of Practical Applications

Table 2.1 illustrates the scope of OIT measurements for assessing quality deficiencies,
processing errors, and other parameters. These are explained in Section 2.2.3 using curves
taken from actual practice.

Application	Example
Thermal stability of a plastic; aging	Comparative measurements allow conclusions to be drawn about degree of stabilization

Application	Example
Presence of stabilizers	Degradation of stabilizer following thermal stress or UV damage
Effectiveness of stabilizers	Melt processing (extrusion, injection molding, welding)
Comparison of aging resistance of different plastics	Damage during service (pipelines, engines)

Table 2.1   Some practical applications of OIT for plastics

2.2   Procedure

2.2.1 In a Nutshell

Specimen preparation   The sampling site is chosen in accordance with the problem (surface/center); as in DSC, the specimen is prepared carefully. For comparative measurements, the specimen surface area should be as constant as possible, as oxidative attack begins at the surface.

Specimen mass   The specimen should weigh as little as possible, because oxidation is usually vigorous. If the stabilizer is not distributed uniformly, it is better to use a heavier specimen; in our experience this is 5–10 mg; according to [8–10] it should be at least 12 mg. The specimen surface area should be as large as possible and above all reproducible. According to [8–10], the specimen thickness is 0.7–1 mm, and the diameter 6 mm.

> **Specimen weight > 5 mg**
> **Specimen surface area as large and**
> **reproducible as possible**

Pan   The DSC pan is open, according to [8–10]; the attacking oxygen has free access to the specimen. An open pan serves as the reference. For copper-induced OIT measurements, copper pans are used.

Purge gas   Oxidation is effected with either oxygen or air as the purge gas. In the static method (T = const.), the specimen is heated in an inert atmosphere.

> **Open pan, oxidative gas atmosphere**

Heating program   **Dynamic OIT**: Measurement starts at room temperature or higher and is continued at a defined heating rate until exothermic decomposition occurs (usually 10 °C/min). The specimen is purged with oxidative gas from the start.

Static OIT: The specimen is heated constantly (usually at 10–20 °C/min) in an inert atmosphere to a temperature far above the melting point, then held isothermally and after a short equilibration time of about 3 min (identical for comparative tests) purged with an oxidative gas (oxygen, air). The gases must be switched as quickly as possible (according to [8–10], within 15 s at most).

> **Dynamic OIT: Oxidation under continuous heating**
> **Static OIT: Oxidation at constant temperature**

Evaluation

Dynamic OIT: The temperature at which an exothermic shift from the baseline occurs (oxidation) is determined.

Static OIT: The time required in an oxidative atmosphere to identify an exothermic shift from the baseline (oxidation) is determined (T = constant).

Interpretation

Static and dynamic OIT studies are to be regarded exclusively as comparative measurements. No predictions can be made about the probable service life of a plastic. Predictions as to crystallinity, melting temperatures, and so forth are of limited value because of poorer heat conduction resulting from missing lids and altered gas atmosphere.

> **OIT measurements are comparative.**
> **Accurate predictions of service life are not possible.**

2.2.2   Influential Factors and Possible Errors During Measurement

2.2.2.1   Specimen Preparation

Plastics are oxidized by the reaction between oxygen and free radicals. These may be generated by heat, irradiation, mechanical shear, and other external factors.

Therefore, as is the case for standard DSC experiments, specimens should be prepared as carefully as possible. For comparative measurements, the contact area of the specimens should be as large as possible and they should have the same geometry .

Because aging and stabilizer degradation do not occur uniformly in parts, specimens should be taken at a representative, predetermined site, for example near the surface or at a point of damage.

Careful, localized sampling is required.
Use specimens with the same surface.

Figure 2.3 illustrates the advantages and disadvantages of various specimen geometries for OIT measurement. Due to the nature of the reaction between oxygen and the specimen the surface area of the specimen is important: a large area yields a clear, easy to evaluate OIT signal. It is best to use the specimen in one piece, where this is possible. Although many small specimen pieces present a large area for the oxygen to attack, it is difficult to reproduce the surface area for comparative measurements, and heat transfer is not the same.

A flat specimen film or a thin specimen offers the best chance of good results because it presents a large surface area relative to its volume for the oxygen to attack and it also ensures good contact with the bottom of the pan. Bulkier specimens offer better heat transfer, but suffer from a greater temperature gradient along the specimen and a small surface area relative to the mass.

The specimen is best removed as a thin wafer and then shaped with a punch. According to [8], specimens should be 0.7–1 mm thick and 6 mm in diameter.

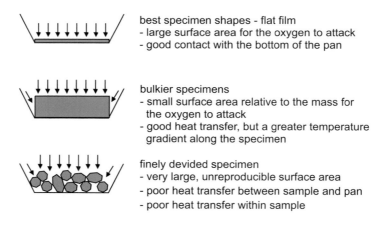

best specimen shapes - flat film
- large surface area for the oxygen to attack
- good contact with the bottom of the pan

bulkier specimens
- small surface area relative to the mass for the oxygen to attack
- good heat transfer, but a greater temperature gradient along the specimen

finely devided specimen
- very large, unreproducible surface area
- poor heat transfer between sample and pan
- poor heat transfer within sample

Fig. 2.3   Advantages and disadvantages of different specimen shapes

Figure 2.4 shows how the **sampling site** affects the results of static OIT measurements. Microtome sections roughly 50 μm thick were taken layer for layer from a 10-year-old PP/EPDM fender, starting from the outside and going down to a depth of roughly 1000 μm.

The static method revealed that the OIT is shortest at the surface of the specimen. This means that the most damage occurred there as a result of extensive degradation of the stabilizer system by environmental effects (UV radiation, water, etc.). The OIT values increase down to a depth of 40 μm. Inside the part, the constant OIT value indicates that the stabilizer system there is still intact.

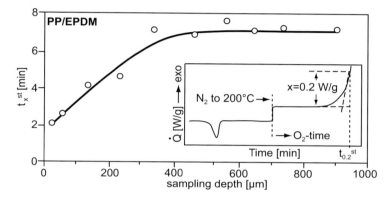

Fig. 2.4   OIT time t_x^{stat} (x = 0.2 W/g) vs. sampling depth in a PP/EPDM fender

Inset: The OIT curve and how it is evaluated

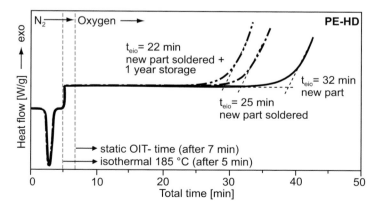

Fig. 2.5   Progressive degradation of damaged PE-HD through aging in a standardized climate

t_{eio} = Extrapolated onset time, specimen mass approx. 14 mg

Another aspect that should not be underestimated during specimen preparation is the **storage time**, that is, the length of time the specimen is stored until the experiment is performed. A plastic begins to oxidize virtually as soon as it is made. Some of the added stabilizer is consumed during processing, post manufacture, and during use. Once oxidation has started, it will continue in storage as well.

Figure 2.5 illustrates this for a new injection-molded PE-HD part that was damaged by soldering in downstream processing. After a year's storage in the laboratory, the soldered specimen exhibited a further shortening of 3 min in OIT value. Therefore, when a plastic is being investigated for its aging characteristics, the time between sampling and measurement should be as short as possible, and the time that elapses before the damage occurs must also be taken into account.

2.2.2.2   Specimen Mass

The reproducibility of OIT measurements is heavily dependent not only on stabilizer distribution, (which can vary in the case of pellets and molded parts from one sampling site to another), but also on the specimen mass.

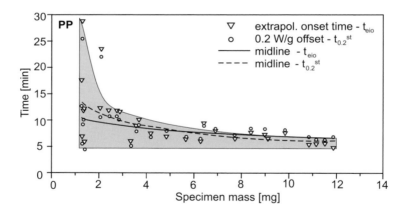

Fig. 2.6   Influence of specimen mass on OIT times, PP document folder,
                holding temperature 200 °C

t_{eio} = *Extrapolated onset time,* $t_{0.2}^{st}$ = *time after baseline offset of 0.2 W/g*

Figure 2.6 is a plot of static OIT readings as a function of specimen mass for PP. The specimens were cut out with a hole punch from a clear PP document folder. Different weights were obtained by stacking several specimens on top of each other and compressing the stack. The curves were evaluated using the extrapolated onset time t_{eio} and a 0.2 W/g baseline offset.

Small specimen masses of 1–3 mg lead to major differences in OIT values. In other words, the reproducibility is extremely poor. This may be due to variations in stabilizer distribution in the PP. Above a specimen mass of approx. 5 mg, this influence can be offset by using several specimens from different yet comparable sampling sites. The results are then more reliable. A specimen mass of at least 12 mg is recommended in [8–10].

A specimen weight > 10 mg increases reliability of the OIT measurement

2.2.2.3   Pans

All standard DSC pans may be used. Aluminum pans are usually adequate, as plastics oxidize below 600 °C.

According to [8–10] the specimen pan is left open in   to facilitate oxygen access. The reference pan is therefore also left open. The specimen should be shaped such that it makes good contact with the bottom of the pan.

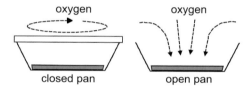

Fig. 2.7   Oxygen attack in closed and open pans

ASTM D 3895 [9] describes the procedure for performing OIT measurements; it uses special copper pans.

2.2.2.4   Purge Gas

In dynamic and static OIT measurements, the inert gases must be highly pure (99.99%) to prevent premature oxidation. Pure oxygen (99,5% extra dry O_2) generally causes a much earlier onset of oxidation than air, which has an oxygen content of approx. 21%, Fig. 2.8.

The purge gas has an even greater influence in the static method, Fig. 2.9. Specimens of PE-HD pellets were each heated to 185 °C in an inert gas atmosphere. After an equilibration time of 2 min, the purge gas was switched to oxygen or to ambient air. Pure oxygen triggered exothermic oxidation after 32 min (total time 39 min; heating time 5 min; equilibration time 2 min = t_{eio} 32 min). When air was used, no oxidation was identifiable even after 50 min had elapsed.

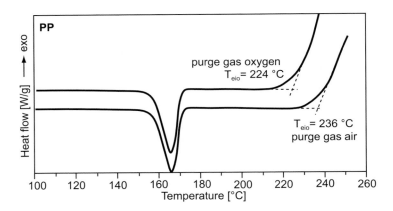

Fig. 2.8   Influence of the oxidative purge gas on the extrapolated onset temperature T_{eio} of PP, dyn. OIT

Specimen weight approx. 10 mg, heating rate 10 °C/min

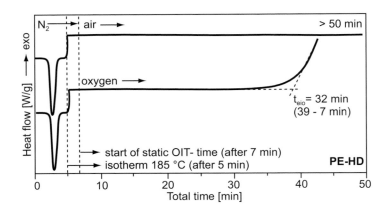

Fig. 2.9   Influence of oxidative purge gas on the extrapolated onset time t_{eio} of PE-HD, static OIT

Specimen weight approx. 10 mg

Although experiments involving air as the oxidative purge gas take longer, this drawback is compensated by the possibility to discriminate more finely between different degrees of stabilization due to the more sluggish oxidation reaction.

2.2.2.5   *Measuring Program*

Dynamic OIT

In dynamic OIT, the measurement is affected primarily by the **heating rate**, Fig. 2.10. Rapid heating shifts the extrapolated onset temperature T_{eio} toward higher temperatures because of thermal gradients and the shorter exposure to oxygen. It also amplifies the heat flux signal, which is directly proportional to the heating rate (see Section 1.1.1).

The end temperature of dynamic OIT measurements depends on the start of decomposition and cannot be predicted accurately. As a rule of thumb, for most plastics it is between 250 °C and 350 °C.

The starting temperature should be lower than the melting point but does not have any bearing on the experimental result.

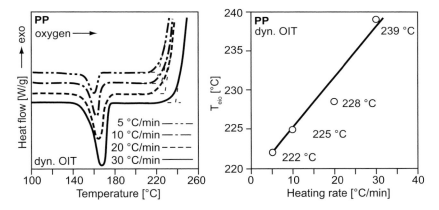

Fig. 2.10   Left: Dynamic OIT curves for PP, measured at different heating rates
                       Right:Plot of extrapolated onset temperature T_{eio} against heating rate

Varying the heating rate may lead to finer discrimination between specimens with different stabilizer content, as illustrated in Fig. 2.11. Old PP specimens (already in use) and new PP specimens were each measured dynamically at a heating rate of 5 °C/min and 20 °C/min.

At the lower heating rate, the extrapolated onset temperature T_{eio} of the old PP is 7 °C lower than that of the new PP.

The higher heating rate of 20 °C/min produces a much greater difference of 21 °C. This may be due to the interaction of, for example, reaction rate, activation energy and incipient thermal degradation.

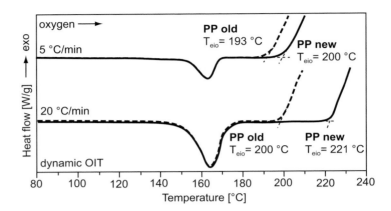

Fig. 2.11   Influence of heating rate on the extrapolated onset temperature T_{eio} of two PP specimens

Specimen weight approx. 10 mg

Static OIT

The rate at which the specimen is heated to the isothermal temperature is of minor importance unless overshooting occurs. The equilibration time for allowing thermal equilibrium to occur at the isothermal temperature in an inert gas atmosphere before the purge gas is changed is approx. 3 min and should always be the same for comparative measurements. If it is too long (e.g., 30 min), thermal decomposition may occur during this time. The purge gas should be switched as quickly as possible. A maximum of 15 s is recommended in [8].

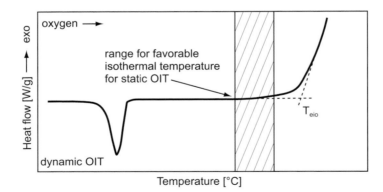

Fig. 2.12   Dynamic OIT measurement for determining a favorable isothermal temperature for static OIT

The crucial experimental parameter is the **isothermal temperature**. Low temperatures make for long experiments. High temperatures cause rapid oxidation and lead to poor reproducibility and poor resolution of OIT times. For every material, a compromise has to be found between an acceptable measuring time and good resolution. A first estimation can be gained by performing a dynamic OIT to determine the extrapolated onset temperature. This curve is used for selecting the isothermal temperature, which is usually just below the extrapolated onset temperature T_{eio}, as illustrated in Fig. 2.12. The temperature window available is relatively small but it must be observed if meaningful OIT times are to be determined. Figure 2.13 illustrates how the value of the holding temperature influences the onset of oxidation of PP. At a holding temperature of 185 °C, oxidation has not even started after 1 h. At 190 °C, it starts after roughly 48 min and at 195 °C, after just 10 min. Because OIT times of between 5 and 60 min are desirable, as they ensure reproducibility and allow the measurement to be performed in an economically acceptable timeframe, an isothermal temperature of 190 °C would appear appropriate.

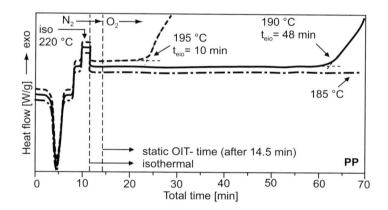

Fig. 2.13   Influence of isothermal temperatures on the extrapolated onset temperature T_{eio} of PP in static OIT

Specimen weight approx. 12 mg

Wellchosen isothermal temperatures lead to OIT times of between 5 and 60 min.

Note: *In our experience, it is often a good idea to heat in an inert atmosphere to a temperature 10–20 °C higher than the isothermal temperature, to hold this for 1–2 minutes and then to return to the desired isothermal temperature. This approach ensures that all the material melts uniformly and that the bottom of the pan is well wetted by the specimen.*

2.2.2.6   Evaluation

As explained in Section 2.1.4, oxidative onset can be determined from OIT curves with the aid of a tangent or a defined baseline shift.

The choice of evaluation method depends on the nature of the problem and the shape of the curve. Where the curves are stable and free of noise, rising sharply in the appropriate range, the results yielded by the two methods are virtually the same, provided that a constant value (frequently 0.2 W/g) is taken for baseline shift. The latter method permits reproducible evaluation of unstable, noisy curves.

Differences between the two evaluation methods arise when slow oxidation reactions yield a flat, exothermic leading edge [12]. This is illustrated in Fig. 2.14 for static OIT measurements on PE-HD specimens aged for different lengths of time. Whereas differences in OIT value are not revealed by the tangent, they are clearly apparent in the 0.2 W/g baseline shift. Because the type of evaluation employed is at the discretion of the tester, it should be recorded meticulously.

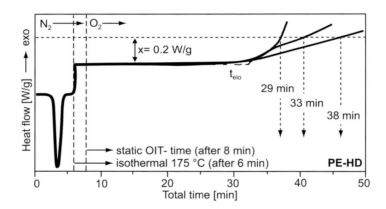

Fig. 2.14   How the evaluation method affects differentiation of PE-HD specimens aged for
               different periods

Record evaluation method: Tangent or baseline shift

*Note: When baseline shift is used in comparative measurements to evaluate dynamic OIT
            at different heating rates, the baseline shift must be correlated with the heating
            rate, for example, 0.1 W/g shift for a heating rate of 10 °C/min, and 0.3 W/g for a
            heating rate of 30 °C/min.*

Reproducibility of OIT measurements
The results of OIT may show great variation, making it difficult sometimes to distinguish between inaccurate measurements and actual effects, for example, different degrees of stabilization.

When all other influences have been taken into consideration, the static method can yield OIT times reproducible to +/– 1 min. The specimen should be inspected after every measurement. Where the degree of wetting is significantly different, the measurements should be rejected.

The dynamic method is not as prone to this problem, as the temperature is continually increased and, at the start of oxidation, is higher than the isothermal temperature. This promotes complete wetting of the bottom of the pan. Dynamic OIT temperatures are therefore usually reproducible to +/– 1 °C.

Because OIT measurements are comparative by nature, the specimens from any one series must be as similar as possible, particularly with regard to surface area of specimens after melting. A reproducible surface area is very difficult to obtain for specimens taken from molded parts and pellets. In such cases, the specimen masses should at the very least be the same. Despite these measures, the pan may still be wetted to different extents and therefore oxygen attack will vary in intensity.

> **Static OIT: OIT times reproducible to +/– 1 min**
> **Dynamic OIT: OIT temperatures reproducible to +/– 1 °C**

Figure 2.15 illustrates the differences in wetting by PP pellets, as observed after a static OIT measurement. The different films formed led to large differences in OIT times (in this case, as much as 22 min).

The cause of this difference in wetting behavior is the change in polarity, viscosity, and density of the polymer during oxidation. Generally, the polar component of the specimen's surface tension increases and, if the melt viscosity is low enough, the surface of the metal pan becomes wetted. This process may be hampered, for example, by release of volatile oxidation products or crosslinking [3]. Contamination of the pan surface also affects wettability.

A continuous, well-melted film with large surface area is conducive to good reproducibility.

Note: *Effects 1 and 2 (Fig. 2.15) predominate in the case of PP and Ziegler PE-HD (Ti type). Effect 3 predominates in the case of PE-HD-Cr (Phillips type) owing to crosslinking during oxidation [3].*

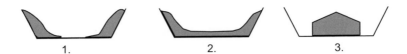

(1)	The melt film creeps up the sides of the pan and tears at the bottom; large, noncontinuous surface area Short OIT times: here, approx. 26 min	Short OIT time, poor reproducibility
(2)	The melt film creeps up the sides of the pan but remains intact on the bottom; large, continuous surface area; medium OIT times: here, approx. 33 m	Medium OIT time, good reproducibility
(3)	The specimen more or less retains its shape; small surface area; long OIT times: here, approx. 48 min	Long OIT time, poor reproducibility

Fig. 2.15   Different types of pan wetting by the specimen

Examine the melted film in the pan after the measurement; a large, continuous surface area is optimum.

Risks of OIT measurements
In Section 2.1.1, we referred to the problems surrounding OIT measurements, particularly their limited ability to predict the long-term behavior of plastics.

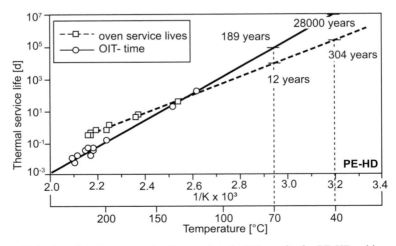

Fig. 2.16   Arrhenius plot of oven service lives and static OIT results for PE-HD cable insulation [13]

Figure 2.16 is an extrapolated plot of service life against temperature for PE-HD telephone cable insulation [13]. Both oven aging studies and OIT times have been plotted. As may be seen, extrapolation of the OIT times overestimates the service life relative to the oven service times. Even at only 70 °C, the service life is more than 10 times as long and this difference becomes wider at low temperatures.

Again, we point out at this stage that it is not possible to distinguish between or compare different stabilizers (see Section 2.1.1) owing to their different modes of action.

Extrapolation of OIT times is not suitable for predicting service life. Comparisons are possible only if the same stabilizer system is used.

OIT values are used to characterize stabilizer degradation. If the stabilizer in a plastic is already totally degraded, further damage reduces the OIT time or temperature because of chain degradation in the polymer. OIT cannot be used to determine the cause of the change in values. Complementary studies, such as viscosity measurements or chromatographic analyses, are recommended.

2.2.3 Real-Life Examples

Section 2.1.5 dealt briefly with the practical applications of OIT measurements.

2.2.3.1 Comparison of "Static" and "Dynamic" Methods

To characterize the differences between the two OIT methods, specimens of PE-HD with low and high stabilizer levels were investigated by the two methods. Figure 2.17 shows the dynamic OIT curves. Two differently stabilized polyethylenes differ in oxidative onset temperature T_{eio} by 5 °C. As the difference is comparatively small, it would not be possible to identify reliably gradations of stabilization between these two extremes. However, the measurement takes only approx. 20 min to perform and thus yields a result in a very short time.

Static OIT measurements performed on these specimens at 175 °C revealed a difference of 20 min in extrapolated onset time T_{eio}, Fig. 2.18. This offers much better resolution for further identifying close gradations in degree of stabilization.

However, greater sensitivity toward specimens with vastly different stabilizer contents can extend the measuring time to beyond the recommended timeframe of 3–60 min.

Dynamic OIT: Faster results
Static OIT: Better resolution

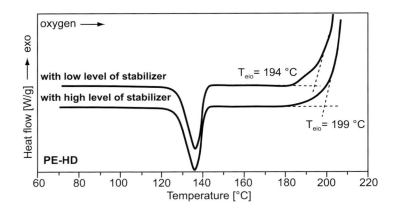

Fig. 2.17   Dynamic OIT of PE-HD with low and high levels of stabilizer

T_{eio} = *Extrapolated onset temperature, oxygen purge gas, specimen mass
approx. 8 mg, heating rate 10 °C/min*

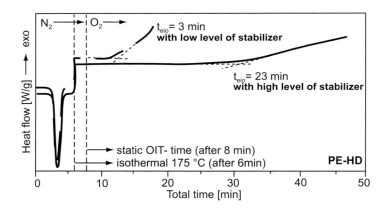

Fig. 2.18   Static OIT of PE-HD specimens with low and high stabilizer contents

t_{eio} = *Extrapolated onset time, purge gas after 8 min: oxygen,
specimen mass approx. 8 mg*

2.2.3.2   Influence of the Processing Parameters

Processing parameters, such as temperature and dwell time, already consume some of the
stabilizers intended for processing. This need not have a negative influence on subsequent

properties of the parts. However, should it prove necessary to check different parts for uniform processing conditions as soon as they have been made, OIT can prove to be a useful tool.

Figure 2.19 uses the example of PS to illustrate how different processing conditions affect subsequent dynamic OIT of the material. Both "hard" and "soft" grades have a lower OIT $T_{0.2 \ W/g}$ value than virgin material. The values can be used to characterize processing in terms of "hard" and "soft" machine settings. Viscosity measurements confirmed that consumption of stabilizer was involved here, and not material degradation.

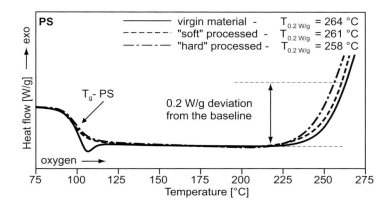

Fig. 2.19   Dynamic OIT of PS at different processing conditions,
              mould temperature: 260 °C (hard) -210 °C (soft)

$T_{0.2}^{dy}$ = Temperature after 0.2 W/g deviation from the baseline,

purge gas: oxygen, specimen mass approx. 10 mg

2.2.3.3   Stabilizer Degradation Through Repeated Processing

Thermoplastics are exposed to elevated temperatures even as they are being pelletized, injection molded, extruded, and processed, for example, welded. These temperatures far exceed the glass transition of amorphous thermoplastics and are also above the melting point of semicrystalline thermoplastics. Consequently, the plastics are vulnerable to thermal, mechanical and oxidative damage.

As previously discussed, although process stabilizers counteract possible degradation, they are also consumed to some extent during processing. Repeated processing, such as recycling, promotes degradation and can cause damage to the material. The effect of repeated processing on the static OIT curve of PP is shown in Fig. 2.20.

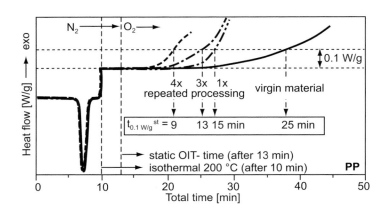

Fig. 2.20   Stabilizer degradation on repeated processing of PP, static OIT

$t_{0.1}^{st}$ = Oxidation time after 0.1 W/g shift, purge gas after 13 min: oxygen

The steady reduction in OIT time indicates degradation of stabilizer and possibly of polymer chains. In this case, a baseline shift of 0.1 W/g was evaluated, as the tangent method appeared inappropriate owing to the leading edge.

Dynamic OIT was used to observe degradation of stabilizer caused by repeated processing of a mixture of ABS and PC. The OIT temperatures fall with the number of processing stages because in this case, too, the stabilizer system is becoming less effective.

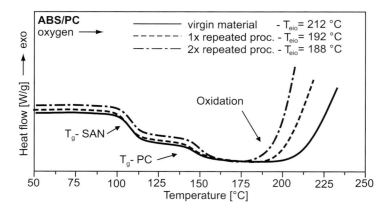

Fig. 2.21   Influence of repeated processing on ABS/PC mixtures, dynamic OIT

T_{eio} = Extrapolated onset temperature, heating rate 10 °C/min, specimen mass approx. 14 mg, purge gas: oxygen

Repeated processing can reduce stabilizer effectiveness.

2.2.3.4 Stabilizer Degradation Due to Effects of Media

Aside from thermooxidative stress, the media contacting the plastic may impair the effectiveness of the stabilizer by dissolving some or all of it out of the plastic. This effect was studied on specimens taken from the inside, outside, and center of the cross section of a PP water pipe, Fig. 2.21.

Static OIT revealed much lower extrapolated onset times for oxidation T_{eio} at the pipe's inner and outer surfaces relative to the center. Obviously the stabilizer was degraded from the outside in. This is not purely a thermal effect, as the low wall thickness of 3 mm would suggest uniform postcuring. On the inside, the stabilizer may have been extracted by the water coursing through the pipe, whereas the outside of the pipe was exposed to oxidative attack by atmospheric oxygen.

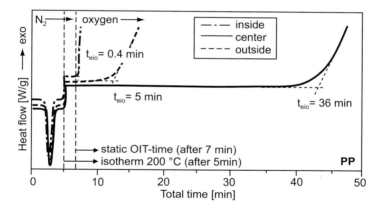

Fig. 2.22   Comparison of extrapolated onset times T_{eio} for a PP water pipe, inside, outside and center, static OIT

Specimen weight approx. 11 mg

2.2.3.5 Results of Round-Robin Trials

Data from various round-robin trials clearly showed that determining the oxidative induction time by the classical static variant (OIT) yields readings that exhibit considerable scatter, especially at very low OIT values [14]. The high values for the standard deviation of repeatability and comparability also revealed that the information yielded by OIT measurements for, for example, quality control or predicting the service lives of polyolefin parts, needs to be regarded with caution.

Poor comparability at low static OIT times

Very low OIT values (materials containing little or no stabilizer), especially, seem to benefit greatly from dynamic (OIT*). The trials additionally revealed that the ability to differentiate between individual samples falls rapidly as the OIT* increases. Here, static OIT measurements, through a lowering of the temperature of the isothermal phase (T < 210 °C) or a reduction in the oxygen content of the purge gas in the measuring chamber, might make it possible to reduce scatter among the readings and to increase the ease with which similar samples may be distinguished [14].

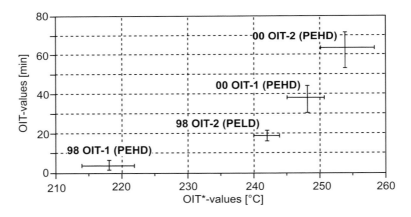

Fig. 2. 23   Plot of pairs of OIT/OIT* values for samples measured in round-robin trials in 1998 and 2000, showing the corresponding limits of comparability (R) in the form of error ranges [14]

2.3   References

[1]   Kramer, E.,   Thermo-Oxidative Degradation of Polyolefins
        Koppelmann, J.   Observed by Isothermal Long-Term DTA
                                         J. Polym. Sci. Eng. 27 (1987) p. 945

[2]   Zweifel, H.   Stabilization of Polymeric Materials
                                            Springer-Verlag, Berlin Heidelberg 1998

[3]   Kramer, E.   Personal Information, April 1998

[4]   Pauquet, J.R.,   Limitations and Applications of Oxidative Induction
       Todesco, R.V.,   Time (OIT) to Quality Control of Polyolefins
       Drake, W.O.   Presented at the 42nd International Wire & Cable
                                          Symposium, November 1993

[5]   Kramer, E.,   Oxidation Stability of Cross-Linked Polyethylene,
       Koppelmann, J.,   Isothermal DTA-Method
       Dobrowsky, J.   J. Thermal Analysis 35 (1989) p. 443

[6]   Audouin, L.,   16th Intern. Conference on Advances in the Stabilization
       Langlois, V.,   and Degradation of Polymers, Conference Proceedings,
       Verdu, J.   Ed. Patsis, A.V., CH-Luzern 1994, p. 11

[7]   Zweifel, H.   Degradation and Stabilization of Polyolefins During
                                          Melt Processing, 13th Intern. Conference on Advances in
                                          the Stabilization and Degradation of Polymers,
                                          Conference Proceedings, Ed Patsis, A.V.,
                                          CH-Luzern 1991, p. 203

[8]   N.N.   DS 2131.2 (1982)
                                          Dansk Standard, Pipes, Fittings and Joints of PE
                                          Type PEM and PEH for Buried Purge gas Pipelines,
                                          p. 10

[9]   N.N.   ASTM D 3895 (2003)
                                          Standard Test Method for Oxidative Induction Time
                                          of Polyolefins by Differential Scanning Calorimetry

[10]   N.N.   ISO 11357-6 (2002)
                                          Plastics – Differential Scanning Calorimetry (DSC) –
                                          Part 6: Determination of Oxidation Induction Time

[11]   N.N.   ISO 11357-1 (1997)
                                          Plastics – Differential Scanning Calorimetry (DSC) –
                                          Part 1: General Principles

[12]   Knappe, S.   Thermische Analyse in der Qualitätssicherung
                                          Kunststoffe 82 (1992) 10, pp. 993–998

[13]   Leu, K.W.   Aging Behavior of Thermoplastic Polymer Materials
                                        Group of Armaments Services, AC-Laboratorium,
                                        CH-3700 Spiez, 1995

[14]   Schmid, M.,   Interlaboratory Tests on Polymers by DSC:
       Affolter, S.   Determination and Comparison of Oxidation Induction
                                        Time (OIT) and Oxidation Induction Temperature(OIT*)
                                        Polymer Testing 22 (2002), pp. 419–428

Other references:

[15]   Ehrenstein, G.W.,   Thermische Einsatzgrenzen von Technischen
       Pongratz, S.   Kunststoffbauteilen
                                        Springer-VDI-Verlag GmbH, Düsseldorf 1998

[16]   Widmann, G.,   Thermoanalyse
       Riesen R.   Hüthig Buch Verlag GmbH, Heidelberg 1990

[17]   Schmutz, Th.,   Oxidation Induction Time (OIT) – a Tool to
       Kramer, E.,   Characterize the Thermo-Oxidative Degradation of a
       Zweifel, H.   PE-MD Pipe Resin
                                        Presented at the Plastics Pipes IX organized by the
                                        Institute of Materials, Edinburgh, Scotland, UK, 1995

[18]   Kreiter, J.   Mit Thermoanalyse Stabilisierungszustand untersuchen
                                        Plastverarbeiter 41 (1990) 7, p. 60–64

[19]   Fearon, P. K.,   DSC Combined with Chemiluminescence for Studying
       Bigger, S. W.,   Polymer Oxidation
       Billingham, N. C.   Journal of Thermal Analysis and Calorimetry 76 (2004),
                                        pp. 75–83

[20]   Riga, A. T.,   Oxidative Behavior of Materials by Thermal Analytical
       Patterson, G. H.   Techniques
                                        ASTM, West Conshohocken, 1997

[21]   N.N.   ASTM E 2009 (2002)
                                        Standard Test Method for Oxidation Onset Temperatrure
                                        of Hydrocarbons by Differential Scanning Calorimeters

[22]   N.N.   ASTM D 5885 (1997)
                                        Standard Test Method for Oxidative Induction Time of
                                        Polyolefin Geosynthetics by High-Pressure
                                        Differential Scanning Calorimetry

3   Thermogravimetry (TG)

3.1   Principles of TG

3.1.1   Introduction

Thermogravimetry (TG) is used to measure the mass or change in mass of a sample as a function of temperature or time or both. Changes of mass occur during sublimation, evaporation, decomposition, and chemical reaction, magnetic or electrical transformations.

Standards covering TG include ISO 11358 [1], ASTM E 1131 [2] and DIN 51 006 [3]. To avoid confusion with T_g, the abbreviation for the glass-transition temperature, TG analysis is often abbreviated to TGA.

> **Changes of mass as a function of temperature and/or time**

Crucial factors are the choice of purge gas and the conditions present in the specimen chamber. The purge gases consist of inert or oxidizing gases, such as nitrogen, helium, argon and oxygen, or air (in isolated cases, the measurement is also performed in a vacuum). The extent of heat transfer to the specimen depends on the gas flow rate.

3.1.2   Measuring Principle

The deflection of a beam carrying the specimen is held constant by means of an electro-magnetic force feed back system. This compensation signal is used to determine the mass of the specimen via the force needed to maintain the beam in a horizontal position. Figure 3.1 shows a horizontal thermobalance. Vertical TG instruments are also in use.

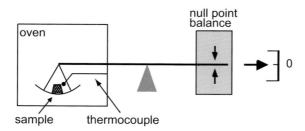

Fig. 3.1   Schematic diagram of a horizontal thermobalance

In addition to standard thermobalances, **simultaneous thermobalances** that measure the mass (TG signal) and temperature difference (DTA or heat flux DSC) of the specimen as it is heated. Thus, the TG curve can be used to describe directly an endothermic process associated with loss of mass (evaporation) and without loss of mass (melting).

Many instruments also permit quantitative information to be ascertained, for example, about heat of fusion or crosslinking enthalpies. However, the less favorable characteristics of these instruments relative to those of a dedicated DSC instrument must be borne in mind.

Combinations of thermoanalytical balances and FTIR or mass spectrometers are also used in polymer analysis. Such combinations are always an advantage when substances are identified by methods involving a certain loss of mass. The underlying principle is that the gaseous components generated during heating in the thermobalance are transferred by a constant gas stream into another test chamber.

In a TGA/FTIR combination, this is performed by a transfer line, which, like the test chamber, is heated to prevent condensation from occurring. Interpretation of the results is a matter of some experience, especially if thermal degradation generates several components simultaneously.

A TGA/MS combination is advisable for gases that are IR-inactive, such as oxygen and nitrogen. The mechanism for combining TG and MS plays a crucial role as the pressure on the TG side has to be converted into a vacuum in the mass spectrometer.

3.1.3 Procedure and Influential Factors

Factors influencing the instrument and the specimen are:

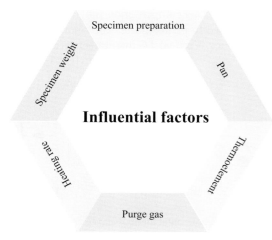

The steps involved in TG are:

- Prepare the specimen preparation.
- Adjust purge gas flow.
- Tare the balance.
- Load specimen input / automatically weigh the specimen mass.
- Select the appropriate heating program.

The influencing factors and possible sources of error during the measurement are explained in detail in Section 3.2.2 with the aid of real-life examples.

3.1.4 Evaluation

In TG, the change of mass of a specimen is measured either absolutely in milligrams or relatively as a percentage of the starting mass, and plotted against temperature or time. Plastics may change mass in one or more steps.

Changes of mass may involve more than one step.

3.1.4.1 *Single-Step Change of Mass*

Figure 3.2 illustrates how the characteristic temperatures involved in a single-step loss of mass are determined in accordance with ISO 11358 [1].

From the TG curve, the points A, B, and C, obtained by means of tangents, and the corresponding temperatures T_A (starting), T_B (end), and T_C (midpoint) are determined. In a masstime plot, the times t_A, t_B, and t_C are evaluated.

The percentage loss of mass M_L is calculated from the masses m_s (at the start, before heating) and m_f (at the end temperature T_B) using the following equation:

$$M_L = \frac{m_s - m_f}{m_s} \times 100 \quad [\%]$$

If an **increase of mass** M_G occurs, for example, during oxidation, Fig. 3.13, this is calculated from the following equation using the maximum mass m_{max} and the starting mass m_s as follows:

$$M_G = \frac{m_{max} - m_s}{m_s} \times 100 \quad [\%]$$

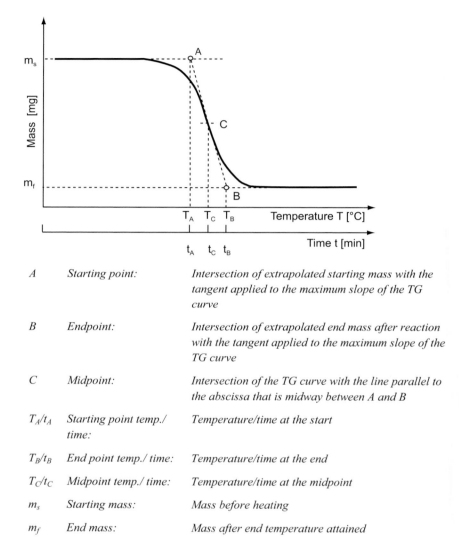

A	*Starting point:*	*Intersection of extrapolated starting mass with the tangent applied to the maximum slope of the TG curve*
B	*Endpoint:*	*Intersection of extrapolated end mass after reaction with the tangent applied to the maximum slope of the TG curve*
C	*Midpoint:*	*Intersection of the TG curve with the line parallel to the abscissa that is midway between A and B*
T_A/t_A	*Starting point temp./ time:*	*Temperature/time at the start*
T_B/t_B	*End point temp./ time:*	*Temperature/time at the end*
T_C/t_C	*Midpoint temp./ time:*	*Temperature/time at the midpoint*
m_s	*Starting mass:*	*Mass before heating*
m_f	*End mass:*	*Mass after end temperature attained*

Fig. 3.2   Evaluation of a typical curve with single-step loss of mass as set out in ISO 11358 [1], TG curve

Table 3.1 lists the characteristics of single-step loss of mass, and the terms by which they are described in different standards.

ISO 11358 [1]	DIN 51006 [3]	Unit
T_A – Onset temperature	T_i – Initial temperature	[°C]
t_A – Onset time	t_i – Initial time	[min]
T_B – End temperature	T_f – Final temperature	[°C]
t_B – End time	t_f – Final time	[min]
T_C – Midpoint temperature	–	[°C]
t_C – Midpoint time	–	[min]
m_s – Starting mass	m_i – Mass at T_i	[mg]
m_f – Final mass	m_f – Mass at T_f	[mg]

Table 3.1   Terms describing characteristics of a TG curve

3.1.4.2   Multistep Change of Mass

In **multi-step losses of mass**, the points to be determined are described with indices A_1, B_1, C_1; A_2, B_2, C_2, etc. and assigned to the corresponding temperatures and times, Fig. 3.3. The resultant temperatures and times follow the same notation, for example, T_{A1}, T_{B1}, T_{C1} and t_{A1}, t_{B1}, t_{C1}.

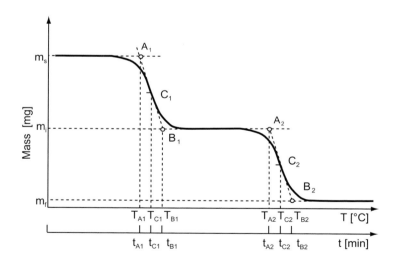

Fig. 3.3   Evaluation of a typical curve for two-step loss of mass, adapted from ISO 11358 [1], TG curve; for explanations, Fig. 3.2

In addition to the starting and final masses m_s and m_f, the mass m_i intermediate between the two losses of mass is determined. To characterize the loss of mass, the differences between these masses are recorded in each case. The first loss of mass M_{L1} and each subsequent loss of mass M_{L2}... are calculated from the following equations:

$$M_{L1} = \frac{m_s - m_i}{m_s} \times 100 \quad [\%]$$

$$M_{L2,..} = \frac{m_i - m_f}{m_s} \times 100 \quad [\%]$$

As shown in Fig. 3.4, multistep TG curves often do not have a section of curve over which the mass remains constant. This is due to a close succession of overlapping changes of mass. In such cases, m_i, the midpoint between m_{B1} and m_{A2}, is determined.

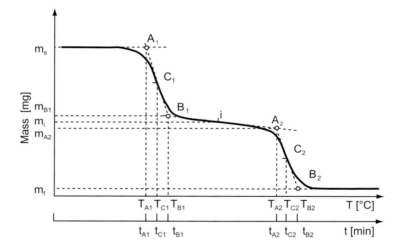

Fig. 3.4   Evaluation of a multistep loss of mass where the mass does not remain constant between the various steps [1]

m_i = Midpoint between m_{B1} and m_{A2}; for explanations, Fig. 3.2

Where this would not be useful either, the time derivative signal dm/dt provides further information. Figure 3.5 is an example of a two-step change of mass. Aside from the TG curve, it contains the differential DTG curve, which can be used to establish m_i, the smallest value on the curve between the two steps.

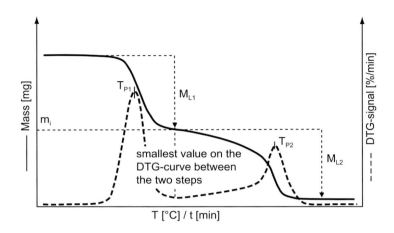

Fig. 3.5   Using the DTG signal to evaluate two steps in close succession

M_{L1}, M_{L2} = Loss of mass, T_{p1}, T_{p2} = peak temperature on the DTG curve

The peak maximum T_p on the DTG curve represents the inflection point of the TG curve and thus the temperature of maximum rate of change of mass. This value is often used in practice for comparison purposes because evaluation is simple and reproducible.

As with other thermoanalytical tests, the evaluated temperatures should be recorded only to the nearest 1 °C owing to the number of factors that can influence them. The loss of mass for a single-step decomposition may be confidently recorded to +/- 0.1%.

Record characteristic temperatures to the nearest 1 °C.

3.1.4.3   Test Report

ISO 11358 provides useful information on how to compile a complete test report that describes all experimental parameters and specimen details.

Where appropriate, the test report should contain the following information [1]:

- Reference to the standards employed.
- All information necessary for complete identification of the material analyzed.
- Size and shape of the test specimens.
- Mass of test specimens.
- Pretreatment of test specimens.

- Type of instrument used.
- Size of specimen container and constituent material.
- Type of temperature sensor employed and its position (inside or outside the specimen container).
- Gas and gas-flow rate.
- Rate of temperature increase (programmed heating), or test temperature in the case of isothermal methods.
- Substance used for temperature calibration.
- Loss of mass or increase of mass or both.
- Residue.
- Temperatures at which changes of mass occur.
- Every observation concerning the instrument, test conditions, or specimen behaviour.
- Date of the test.
- TG curve.

3.1.5 Calibration

Calibrations must be performed under the same conditions as the actual measurement; that is, the heating rate, purge gas flow, rate and position of the thermocouple must all be identical.

The thermobalance has to be corrected for buoyancy due to temperature effects. Furthermore, a calibration is needed for the mass signal or change of mass signal. Finally, as in DSC, the temperature has to be calibrated.

3.1.5.1   Buoyancy Correction

Buoyancy is caused by a change of purge gas density as the temperature changes, as well as by aerodynamic friction between the purge gas at the hangdown wire and the specimen [4]. It is corrected by subtracting a curve produced by a blank measurement (without specimen) from the TG curve of the specimen, under the proviso that all other experimental parameters are equal.

3.1.5.2   Mass Calibration

The thermobalance must be calibrated with calibrated weights having masses between 10 mg and 100 mg; to avoid buoyancy effects and turbulence, no purge gas is passed through the balance during this calibration [1].

3.1.5.3   Temperature Calibration

Calibration must be performed in the same conditions for purge gas, gas flow rate, and heating rate that will be employed for actual test specimens [1]. The "true" specimen temperature is calibrated using the **Curie temperatures** of ferromagnetic substances. When such a substance is placed in a thermobalance and exposed to a magnetic field, it experiences an additional force that is displayed as the apparent change of mass. When heated, the substance loses its ferromagnetic properties at the Curie temperature, with the result that the apparent change of mass disappears again. The displayed temperatures are corrected to the actual tabulated Curie temperatures (see below).

Reference material	T_A [°C]	T_C [°C]	T_B [°C]
Permanorm 3	253	259	267
Nickel	351	353	355
Mumetall	378	382	386
Permanorm 5	451	455	458
Tratoperm	749	750	751

Table 3.2   Calibration substances for magnetic phase transformations [1]

In ISO 11358 [1], this calibration is performed with the aid of two or more standard references that have Curie temperatures in the range under consideration. The procedure to be adopted for a calibration of this kind is discussed in more detail in [1, 5–7].

The disadvantage of the method is that tabulated Curie transformation temperatures are simply mean experimental values with a relatively high degree of scatter [5].

A further method for calibrating the temperature uses the **melting points of pure metals** (indium, lead, zinc, aluminum, silver, gold). It has not yet achieved the status of a standard.

One or more metal bodies are attached to the specimen holder in such a way that they drop as they melt. The best results are obtained by attaching a small weight to the metal to amplify the effect. The value T_A, that is, the intersection of the extrapolated starting mass with the tangent to the abrupt slope, serves to calibrate the temperature. Figure 3.6 illustrates the TG curves of indium and zinc.

Experimental parameter = Calibration parameter

Note: *In our experience, the role of calibrations should not be overexaggerated. Calibration to the nearest tenth or hundredth of a degree is unrealistic owing to the number of factors affecting TG.*

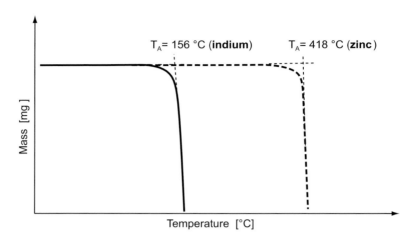

Fig. 3.6   TG curve of indium and zinc for calibrating the temperature by means of dropping
                at the melting point

                Heating rate 10 °C/min, purge gas: nitrogen, T_A = starting temperature (calibrated)

3.1.6 Overview of Practical Applications

Table 3.3 shows which TG characteristics may be used to assess quality deficiencies, compositions of mixtures, processing errors and other parameters. These will be explained in more detail in Section 3.2.3 with the aid of curves taken from real-life examples.

Application	Characteristic	Example
Quantification of material composition	$M_{L1,2}$	Plasticizer and plastic fractions, e.g., in rubber blends, volatile components
Filler content	$M_{L1,2}$	Determination of the content of carbon black, chalk, glass-fibers, or other inorganic fillers
Decomposition behavior (defined atmosphere)	T_A, T_C, T_B, M_L	Start and progress of decomposition at certain temperatures; quantification of decomposition products
Decomposition kinetics	t values	Decomposition rate
Drying time and temperature	T_A, t_A, T_B, t_B	Time and temperature at which the solvent or water present has been completely been removed from, e.g., paint

Application	Characteristic	Example
Moisture content, evaporation of low molecularweight components	M_L	Content of water and volatile substances

Table 3.3   Practical applications of TG measurements on plastics, along with the relevant
             characteristics.

3.2   Procedure

3.2.1 In a Nutshell

Specimen preparation   The specimen must be prepared and removed carefully, especially when volatile low molecular weight components are to be determined.

Specimen mass   The minimum specimen mass per instrument is approx. 1 mg. ISO 11358 recommends a minimum specimen mass of 10 mg [1]. Greater specimen masses are limited to approx. 100 mg by the volume of the specimen pan, and temperature gradients must be expected within the specimen.

> **Specimen weight: 10–100 mg**
> **Clean the pan thoroughly between measurements.**

Pan   Specimen pans are generally of platinum or ceramic, but may also be aluminum (< 600 °C). They may be reused but must be cleaned thoroughly after every measurement.

Purge gas   Both inert (O_2 < 0.001%) and oxidative purge gases are used. Water content of purge gas: < 0.001 wt.%.

Heating program   Dynamic measurements are performed at defined heating rates of between 0.5 and 20 °C/min. Isothermal measurements are conducted as a function of time. The heating rate should be chosen such that effects occurring during heating are minimized but that overshooting by the oven is avoided (approx. 50 °C/min). Combined heating and holding phases enhance resolution of superposed effects. The final temperature for a plastics analysis lies between 600 and 1000 °C.

Evaluation   Decomposition temperatures and times, and changes of TG signal and mass.

Interpretation   Identification of plastics is generally not possible as a large number of additives may affect the decomposition behavior. Combining TG with other analytical methods (mass spectroscopy, purge gas chromatography, infrared spectroscopy, etc.) will identify the decomposition products. Information can be gained about kinetics processes.

Evaluation of decomposition temperatures/times and change of mass

3.2.2 Influential Factors and Possible Errors During Measurement

3.2.2.1   Specimen Preparation

Specimen preparation can have a significant effect on the results of the TG analysis, particularly if the specimen contains readily volatile constituents. These change the starting mass of the specimen even as preparations are being made to conduct the measurements at room temperature. Figure 3.7 illustrates this effect. It shows the results obtained for solvent-borne correction fluid that was analyzed as a function of time at 23 °C.

Nearly all readily volatile substances have evaporated within approx. 5 min at 23 °C. There is a residue of 65% and the correction fluid is dry.

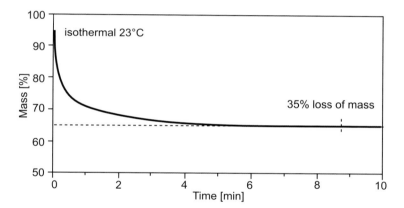

Fig. 3.7   Isothermal TG measurement performed on correction fluid

Temperature 23 °C, purge gas: air

Readily volatile components, for example, solvents or low molecular weight substances, reduce the starting mass m_s.

Similar problems occur with styrene containing reactive resins, as shown by the TG curve produced by an SMC paste, Fig. 3.8.

At room temperature, the styrene contained in the SMC starts escaping as the specimen is being transferred to the pan. When the boiling point of styrene (141 °C) has been reached, loss of mass due to escaping styrene is complete. The measured evaporated fraction of 7% does not, however, correspond to the original styrene fraction in the SMC paste. Because the measurement was performed during heating, the curing reaction that set in polymerized the styrene into the material. This process does not entail loss of weight and thus does not show up in the TG curve.

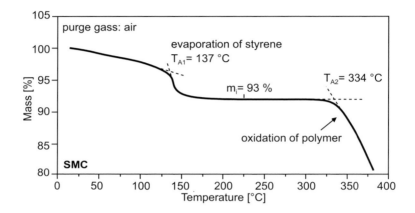

Fig. 3.8   SMC paste, evaporation of styrene with simultaneous (unidentifiable) polymerization

T_{A1}, T_{A2} = Starting temperature of the 1st and 2nd changes of mass
m_i = Mass intermediate between two losses of mass
Sspecimen mass approx.10 mg, heating rate 10 °C/ min, purge gas: air

When filler content is being determined, it should be remembered that fillers are seldom uniformly dispersed throughout a molded part. Fig. 3.9 shows the example of a GRP pressed sheet in which the loss of mass varies with the sampling site, that is, the front or back side of the part. The difference between the losses of mass is 13% and between the filler contents, that is, the residue, is 23 %. The selection and documentation of the sampling site are therefore very important. With fiber-reinforced plastics, sampling is representative of the fiber content only to a certain extent. This is because the relatively small quantity of speci-men includes only fragments. In such cases, the loss on ignition of large specimens should be performed in accordance with EN 60.

Selective sampling is required when filler distribution is nonuniform.

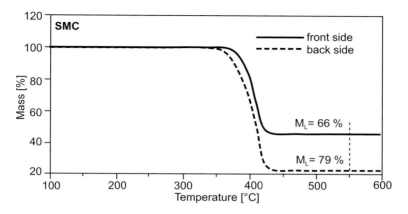

Fig. 3.9   Filler content of front and back side of SMC pressed sheet

M_L = Loss of mass, specimen mass approx. 7 mg, heating rate 10 °C/min, purge gas: nitrogen

3.2.2.2   Specimen Mass

The influence of specimen mass on the starting temperature T_A and the peak maximum of the DTG signal are illustrated in Fig. 3.10, which shows the results for an ultrahigh molecularweight polyethylene (PE-UHMW).

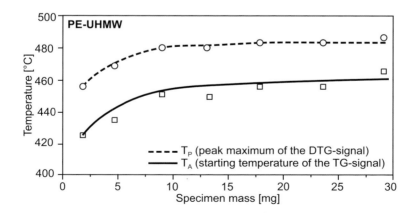

Fig. 3.10   Influence of specimen mass on the results of a TG measurement

Heating rate 10 °C/min, purge gas: nitrogen

In the low specimen mass range, this characteristic temperature initially, stabilizing only at a specimen mass of approx. 10 mg. This is probably due to the fact that small specimens have a larger area-to-volume ratio and the specimen is therefore attacked more rapidly.

For the conditions shown, a minimum specimen mass of 10 mg, as set out in ISO 11358, appears reasonable. Where comparative measurements are to be made at low specimen masses, the weights should be within +/- 1 mg of each other

Suitable specimen mass: > 10 mg
Specimen weights < 10 mg should be within +/– 1 mg of each other
for comparative purposes

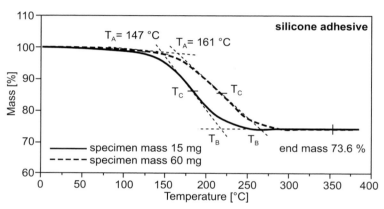

Fig. 3.11   Influence of specimen mass on the characteristic temperatures T_A, T_B and T_C of the TG curve for a silicone adhesive

Heating rate 10 °C/min, purge gas: nitrogen

When the specimen masses are very high, the measurement is influenced by the thermal inertia of the plastic. Figure 3.11 shows the TG curves for two specimens of silicone adhesive of different size. The larger specimen decomposes long after the smaller one. The magnitude of this effect depends on the material properties and the contact area between the specimen and the pan.

3.2.2.3   Pan

The specimen pan is generally made of platinum because temperatures employed for plastics may be as high as approx. 900 °C and mostly involve oxidative atmospheres. The pan can be cleaned thoroughly after every measurement for re-use. It is cleaned by heating at elevated temperatures (600–1200 °C). Residues are then removed with the aid of a brush, glass-fiber pen, or solvent (alcohol or acetone). Where temperatures do not exceed 600 °C, aluminum pans may also be used.

The filler content of DSC specimens can be determined accurately by means of TG. Often, the whole DSC pan (usually aluminum) is placed in the TG balance because the specimen is difficult to remove. The temperature must not exceed 600 °C as, otherwise, the aluminum will melt (660 °C) and react with the platinum pan.

3.2.2.4   Purge Gas

Inert purge gases such as helium, nitrogen, and argon are suitable for determining purely thermal decomposition (pyrolysis). Oxygen and air serve as oxidizing purge gases for determining thermooxidative decomposition (oxidation). The purge gas pressure must be adjusted to ensure that the specimen is completely enveloped. A flow meter is needed for providing precise control.

Vertical thermobalances may suffer from "chimney" effects. These can be suppressed by dividing the inlet purge gas into overflow and underflow purge gas.

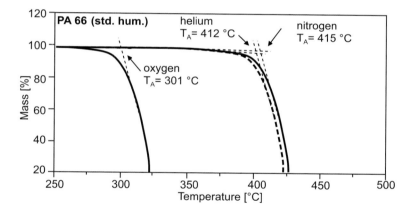

Fig. 3.12   Decomposition behavior of PA 66 (std. hum.) in different atmosphere (oxygen, helium, and nitrogen)

T_A = Starting temperature, heating rate 10 °C/min, specimen mass approx. 8 mg

Note: Because, the 2% water has already evaporated from the PA 66 during heating to 250 °C, the TG curve starts at a mass of 98%.

The nature of the purge gas has a great effect on the experimental results, as the comparison of oxygen, helium and nitrogen in Fig. 3.12 shows. The polyamide decomposes much earlier in an oxidizing atmosphere. Within the range of scatter, no difference is found between helium, and nitrogen (after appropriate calibration).

Inert purge gases, especially, must be highly pure as even traces of oxidizing elements can lead to premature oxidation. ISO 11358 stipulates that the inert purge gas may contain at most 0.001 vol.% oxygen. The system must be flushed properly with the inert purge gas or perhaps evacuated before the start of the measurement.

> **The purge gas determines the type of decomposition.**
> **Inert purge gas: thermal**
> **Oxidizing purge gas: thermooxidative and thermal**

Oxidation of ultrahigh molecularweight PE is accompanied by an initial gain in weight (due to occlusion of O_2), as shown in Fig. 3.13.

While no change of mass is observed at temperatures below 400 °C in nitrogen, the PE decomposes in the presence of oxygen. When oxygen serves as the purge gas, it causes the mass to increase briefly at temperatures around 200 °C due to O_2 occlusion. This is quickly followed by combustion. The increase of mass is even greater when air is used, as decomposition takes place later.

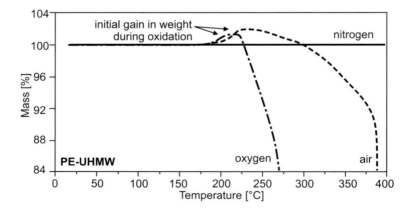

Fig. 3.13   Initial increase of mass during oxidation of PE-UHMW

Heating rate 10 °C/min, specimen mass approx. 11 mg

3.2.2.5 Measuring Program

In **dynamic measurements**, the starting temperature is generally room temperature and the final temperature around 1000 °C. The heating rate has a major affect on the results, as shown in Fig. 3.14 for PP. Doubling the heating rate from 10 to 20 °C/min shifts the starting temperature T_A to a temperature 11 °C higher.

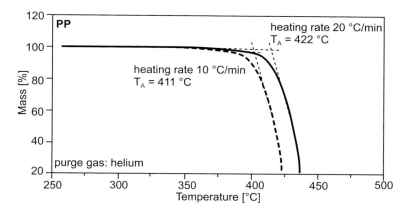

Fig. 3.14   Influence of heating rate, PP, T_A = starting temperature

Specimen weight approx. 6 mg, purge gas: helium

When **isothermal measurements** are being performed, the goal is to reach the test temperature as fast as possible. Although ISO 11358 [1] recommends the fastest possible heating rate (100 °C/min or more) so as to avoid any effects that may arise during heating, we have found that a heating rate of 50 °C/min is better because it avoids excessive "overshooting" (exceeding the measuring temperature briefly due to the thermal inertia of the oven).

In mixtures of plastics, the decomposition temperatures of different components frequently overlap, preventing accurate evaluation. In such cases, combined **heating and isothermal programs** may prove useful, as shown in Fig. 3.15.

A constant heating rate of 10 °C/min up to 600 °C failed to resolve a PP mixture.

The other approach was to heat the specimen at 10 °C/min to 355 °C and hold it at this temperature until no further change of mass occurred. During this phase, which lasted 90 min, the rubber fraction decomposed. The specimen was then heated at 10 °C/min to the final temperature of 600 °C.

The holding temperature must not be too high, as the PP component might decompose if held there long enough.

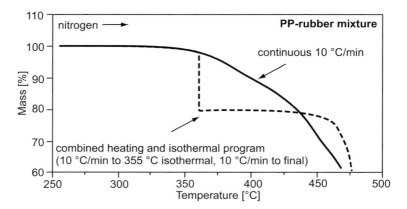

Fig. 3.15   Combined heating-and-holding program vs. continuous heating of a PP-rubber
             mixture

         Specimen weight approx. 10 mg, purge gas: nitrogen

Modern TG instruments allow the heating program to adapt automatically to the decomposition processes. At a predetermined loss of mass, heating is automatically interrupted and
either the temperature is held or the heating rate is reduced, until no further loss of mass
occurs. The original heating rate is then applied.

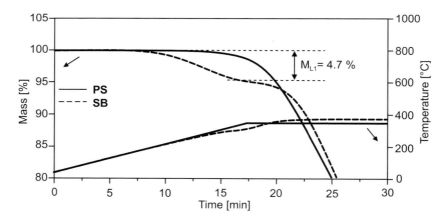

Fig. 3.16   TG (high-res) measurement of PS and SB

         M_{L1}=Mass loss, overall heating rate 20 °C/min (resolution 4, sensitivity 3), specimen
         mass 10 mg, purge gas: nitrogen

This kind of modified heating program can be especially useful for separating mixtures. Figure 3.16 shows the plot for two different polystyrene grades. The automatic heating-rate controller enables the two types to be distinguished. One is a standard grade (PS) and the other is an impact-modified grade (SB) whose rubber content is reflected in the weight loss.

When the end temperature has been reached (600–1000 °C), the furnace must not be opened while it is hot. Otherwise, atmospheric oxygen might diffuse into its pores (furnaces are usually made of ceramic materials). This oxygen would be released in the next heating phase and would lead to spurious oxidation effects even if an inert atmosphere was used, which would falsify the readings.

Cool ceramic ovens to approx. 600 °C in an inert gas atmosphere.

3.2.2.6   Evaluation

Section 3.1.4 discussed how to evaluate the TG signal in terms of characteristic temperatures and changes of mass.

The value of the starting temperature T_A, which is found where the extrapolated curve for the starting mass intersects with the tangent applied to the sharp drop in the TG curve, depends on the slope of the tangent.

Even minor changes in the tangent's slope, which, for example, could arise from applying the tangent at different points in the sharp drop, lead to comparatively large deviations of several degrees in the starting temperature. Where possible, the position of the tangent should therefore be the same for all comparative measurements.

The slope of the extrapolation tangent affects the evaluation.

By contrast, the peak temperature T_p of the differential TG signal may be evaluated to the nearest 1 °C. The loss of mass for a single-step decomposition may be quoted with confidence to +/– 0.1%.

As far as reproducibility is concerned, five consecutive measurements on PBT pellets using the same evaluation method and experimental conditions yielded starting temperatures with a range of scatter of +/– 3 °C and peak temperatures for the DTG signal in the range +/– 2 °C.

To be able to identify the smallest of changes, it is useful to superpose curves or to compare tabulated experimental readings.

3.2.3 Real-Life Examples

3.2.3.1   Carbon Black Content

In plastics technology, carbon black serves as black pigment, providing protection against UV radiation, enhancing heat stability, and imparting electrical conductivity. It is also a common filler in the elastomer sector.

To separate decomposition of carbon black and polymer, different purge gases are employed during measurement. Almost all polymers pyrolyze between 400 °C and 600 °C in an inert atmosphere. Carbon black remains stable in such conditions. It can be oxidatively decomposed afterwards in an oxygen atmosphere, see also ASTM D 6370 [8].

Carbon black content is determined by varying the purge gas.

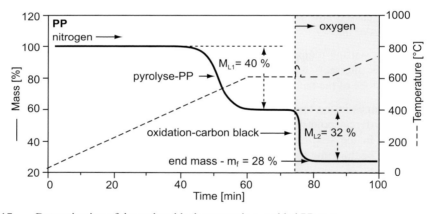

Fig. 3.17   Determination of the carbon black content in a molded PP part

M_{L1} = Loss of mass during pyrolysis (polymer fraction)

M_{L2} =Loss of mass during oxidation (carbon black fraction)

m_f = Mass at final temperature

Heating rate 10 °C/min, specimen mass approx. 12 mg, purge gas change at 600 °C

Figure 3.17 shows the TG curve for PP containing carbon black that was heated under nitrogen at 10 °C/min. The polymer pyrolyzes at approx. 420 °C (T_A). At 600 °C, the heating phase is interrupted; it is followed by an isothermal holding phase during which pyrolysis goes to completion. After 10 minutes, the switchover is made from nitrogen to oxygen and heating continues at a rate of 10 °C/min. The subsequent second step in the curve stems from combustion of the carbon black ($M_{L2} = 32\%$). Immediately after the purge gas change, the temperature curve shows a brief increase in temperature due to this combustion. At the end of the measurement, at a final temperature of 750 °C, an inorganic residue ($m_f = 28\%$) consisting of glass fibers remains.

Note: *When a quantitative determination is being carried out, it should be remembered that carbon black formed during the pyrolysis of the polymer is also oxidized.*

Figure 3.18 shows the oxidation of carbon black formed during pyrolysis. It stems from transparent polycarbonate granules that contain no additives. The plastics decomposes initially in a nitrogen atmosphere. After the changeover to oxygen, the pyrolytic carbon black formed during that decomposition then oxidizes. Polycarbonate generates a particularly high proportion of pyrolytic carbon, namely about 25%.

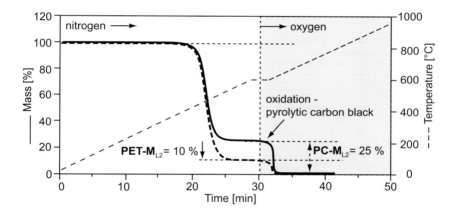

Fig. 3.18   Pyrolytic carbon black produced by different unfilled plastics

Heating rate 20 °C/min, specimen mass approx. 10 mg,
purge gas: nitrogen; oxygen from 600 °C

Aside from PC and PET, the following plastics have substantial contents of pyrolytic carbon black:

PVC: 16.5%	PC: 10–20%	PA 66 : approx. 1.5%
PET: 9–15%	PTFE: > 50%	PEI: approx. 55%
PSU: approx. 45%	PEEK: approx. 54%	

To avoid false interpretation when filler contents are being determined, the content of pyrolytic carbon black must be borne in mind.

Bear pyrolytic carbon black in mind when determining filler content.

3.2.3.2   Calcium Carbonate/Talc Fillers

TG can help to distinguish between the common inorganic fillers chalk (calcium carbonate) and talc. Calcium carbonate decomposes in a nitrogen atmosphere, but talc does not show any loss of mass, Fig. 3.19.

Pure calcium carbonate decomposes into carbon dioxide and calcium oxide:

$$CaCO_3 \leftrightarrow CO_2 \uparrow + \; CaO$$

The release of gaseous CO_2 leads to a 44% loss of mass (M_L). The chalk content of filled plastics can thus be quantitatively determined by applying a conversion factor for the thermogravimetric change of mass due to loss of CO_2 (factor: 100/44, i.e., 2.27).

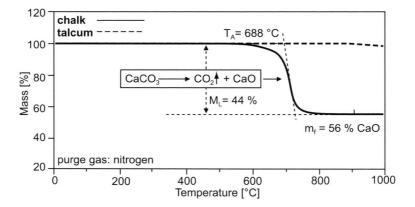

Fig. 3.19   TG measurements for pure chalk and pure talcum

T_A = Starting temperature, M_L =Loss of mass
m_f = Mass at final temperature
Heating rate 10 °C/min, specimen mass approx. 10 mg, purge gas: nitrogen

Figure 3.20 shows how the calcium carbonate fraction is calculated in a filled PP. The PP starts to pyrolyze at a starting temperature T_A of 417 °C. The extent of decomposition is 71%. Subsequent loss of mass is due to escaping CO_2.

Multiplication of the 12.6% loss of mass by the factor 2.27 yields an initial chalk fraction of 29%. The 16.4% residue is calcium oxide [9].

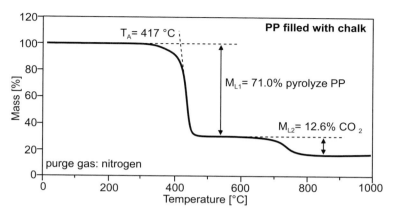

Fig. 3.20   Determination of the chalk fraction in PP [9]

T_A = Starting temperature, M_{L1}, M_{L2} = Loss of mass

Heating rate 10 °C/min, specimen mass 17.7 mg, purge gas: nitrogen

3.2.3.3   Stabilizer Degradation

TG can be used to establish the extent to which the polymer has been stabilized or aged. It can be used to compare polymers, however, only if they contain the same stabilizer system. Unlike OIT (see Section 1.1), which measures the exothermic oxidation reaction associated with oxidation, TG uses loss of mass to reveal the start of decomposition.

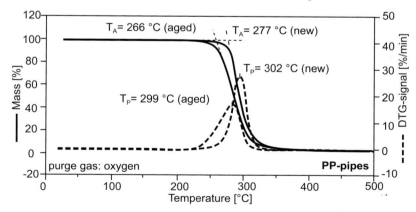

Fig. 3.21   Distinguishing between new and aged PP pipes

T_A = Starting temperature

T_P = Peak temperature of the DTG signal

Heating rate 10 °C/min, specimen mass approx. 10 mg, purge gas: air

Figure 3.21 shows the TG and DTG curves for the decomposition behavior of specimens from new and aged PP pipes. The starting temperature T_A of the aged specimen is 11 °C lower and the peak temperature T_p is 3 °C lower than in the virgin material. There is a marked difference in the shape of the DTG peaks.

3.2.3.4   Repeated Processing

As we have already shown using DSC and OIT measurements, when polymers are repeatedly processed, the stabilizer is consumed or the molecular chains are degraded.

Figure 3.22 shows the pyrolysis of PP processed once and three times. Note that the temperature scale covers only the range 300–450 °C, that is, the differences in decomposition behavior are very slight. For evaluation purposes, therefore, it is advisable to pick a specific figure for loss of mass (e.g., 1, 5, or 10%) to characterize decomposition. In this example, the chosen value of 10% yielded a 10 °C difference for the start of thermal degradation. This approach is recommended when it is difficult to apply tangents to the curve, for example, when decomposition reactions are superposed.

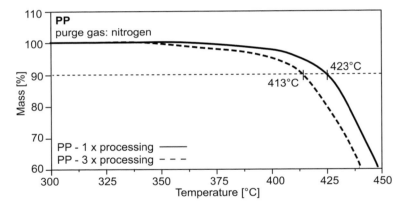

Fig. 3.22   Influence of repeated processing on the decomposition behavior of PP, temperature for 10% loss of mass

Heating rate 10 °C/min, specimen mass approx. 10 mg, purge gas: nitrogen

3.2.3.5   Multistep Decomposition

Pyrolysis of polymers blends or copolymers can also take place in several steps. Figure 3.23 shows this for SEBS.

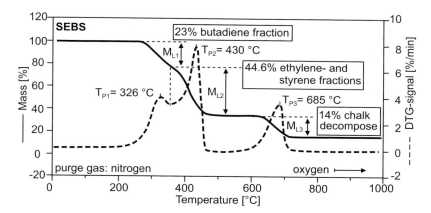

Fig. 3.23   Three-step loss of mass of SEBS

T_P = Peak temperature of the DTG curve
M_L = Loss of mass
Heating rate 10 °C/min, specimen mass 11 mg, purge gas: nitrogen;
oxygen purge gas from 850 °C

The first step involves decomposition of butadiene, leading to a loss of mass M_{L1} of 23%. This is followed by decomposition of the ethylene and styrene fractions with a total mass loss M_{L2} of 44.6%. Just below 700 °C, the chalk filler decomposes into calcium oxide and carbon dioxide (see Section 3.2.3.2). The chalk content is calculated by multiplying the carbon dioxide fraction of 14% by the factor 2.27 to yield a value of 31.8%. After the purge gas is switched from nitrogen to oxygen at 850 °C, no further weight loss is evident. This indicates that the material did not contain any carbon black.

Pure PVC also undergoes multistep decomposition. From 250 °C on, hydrogen chloride (HCl) is released from PVC in a nitrogen atmosphere. This manifests itself in a loss of mass of 63 %. The hydrocarbon backbone then decomposes above 400 °C, Fig. 3.24.

The height of the step at which HCl release occurs can be used to determine the PVC content of plastic blends, as shown in the top right side of Fig. 3.24 for a PE-HD/PVC blend. A mass fraction of 1% PVC gives rise to 0.63% loss of mass, as determined from the curve for 100% PVC. When the PVC fraction is less than 1%, the measurement is no longer reproducible.

In practice, release of HCl is frequently masked by other effects that can falsify the result, for example, the decomposition of low molecularweight additives or aging. Meaningful comparisons can therefore be obtained only where the same grade of PVC has been used for blends and the pure compound, and where comparable history can be assumed – which is seldom the case in practice.

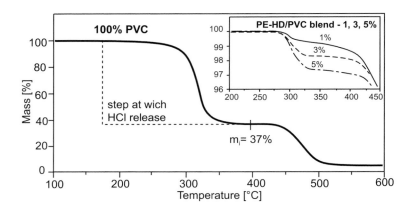

Fig. 3.24   TG measurement of pure PVC and PE-HD/PVC blends (top right); the loss of mass of
                the first decomposition step was evaluated; m_i = mass between two decomposition steps

                Heating rate 10 °C/ min, specimen mass approx. 8 mg, purge gas: nitrogen

3.2.3.6   Media Effects

Media in contact with a plastic can extract stabilizers and so trigger premature chain degradation. They can also dissolve into the plastic.

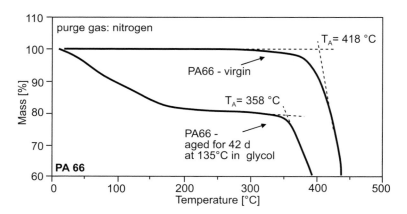

Fig. 3.25   TG measurement performed on PA 66, virgin and aged for 42 days at 135 °C in glycol

                *T_A = Starting temperature, heating rate 10 °C/min, specimen mass approx. 12 mg,
                purge gas: nitrogen*

An injection molded PA 66 part was immersed at 135 °C in pure glycol for 42 days and then compared with an "as made" injection molded part. As shown in Fig. 3.25, major differences were found in the decomposition behavior. From the very start of heating, the aged material lost weight as the glycol, constituting some 20%, evaporated. Once the boiling point of the glycol (180 °C) was exceeded, loss of mass ceased. The rest of the curve differs from the non-aged part in having a lower starting temperature for decomposition – this indicates degradation of the PA 66 or its stabilizers by the glycol. The effect of machine oil on the decomposition of a metallocene-polyethylene (mPE) specimen is shown in Fig. 3.26.

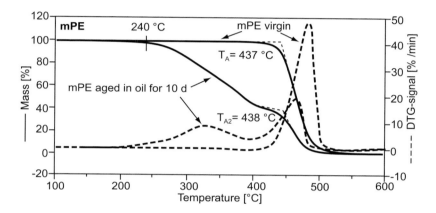

Fig. 3.26   TG analysis of mPE, virgin material, and material aged in oil for 10 days

T_A = Starting temperature, heating rate 20 °C/min, specimen mass approx. 10 mg, purge gas: nitrogen

Note: The specimen must be cleaned thoroughly prior to the experiment (with cotton cloth or cellulose, but never with solvent) to remove any adhering aging medium.

The oil present in the specimen after aging evaporates from the material at roughly 240 °C and above. The resultant loss of mass is 40%. Above 400 °C, the plastic itself decomposes, with no difference discernible between the starting temperatures T_A of the new and the aged materials. It may be concluded from this that the plastic was not damaged by the machine oil.

As with other methods of thermal analysis, TG measurements on their own often do not provide a comprehensive picture of the qualitative or quantitative composition of plastics. When superposition of decomposition processes occur, not even calibration curves will prove to be of much help. Combination with other analytical methods such as liquid and purge gas chromatography and infrared and mass spectroscopy may be of assistance in such cases.

Ambient media can diffuse into and damage plastics

3.2.3.7   Degradation of Polyoxymethylene

TG analyses of POM degradation are performed in an inert nitrogen atmosphere. The initial decomposition temperature (IDT) is defined as the temperature at which a weight loss of 1% occurs. This temperature signifies the start of thermal degradation and, according to [10], is in step with the molecular weight. Figure 3.27 shows the TG curves for a degraded and a nondegraded POM-H.

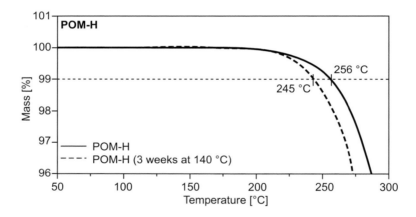

Fig. 3.27   Influence of material degradation on the shape of the TG curve of POM-H

Heating rate 10 °C/min, specimen mass approx. 10 mg, purge gas: nitrogen

The 14 °C difference in IDT is evidence of material degradation. Because measurements of the molecular weight or the solution viscosity of POM are very laborious, TG is the preferred method for characterizing degradation.

3.2.3.8   Results of Round-Robin Trials

The materials employed were polyolefin granules containing 2–3% carbon black (comercial samples and material homogenized by extrusion) and vulcanized elastomers containing more than 30% carbon black. Figure 3.28 shows a typical curve.

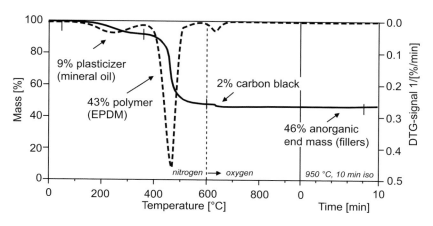

Fig. 3.28   TG and DTG curves for a polymer compound [11]

The following table summarizes the round-robin results for carbon black content, repeatability, and comparability and indicates the specimen form.

Specimen	Form	Carbon black content y [%]	Repeatability r [%]	Comparability R [%]
PE-LD 1	Granules	2.35	0.60	0.69
PE-LD 1	Regranulated	2.34	0.10	0.30
PE-LD 2	Granules	2.33	0.27	0.72
PE-LD 2	Regranulated	2.29	0.07	0.32
TPE	Granules	2.84	0.18	0.81
TPE	Regranulated	2.86	0.15	0.66
IIR vulcanized		41.78	0.40	2.15
SBR/BR/NR blend vulcanized		33.84	0.48	1.50

Table 3.4   Round-robin determination of the carbon-black content y, repeatability r (same laboratory), and comparability R (different laboratory) for different polymer compounds

The following insights were gained from these trials:

Homogenization of the PE-LD specimens substantially reduced the repeatability r and the comparability R. Caution must therefore be exercised when carbon black contents of raw materials are being determined for QC purposes.

High carbon black contents can be expected to have a relative repeatability of 2%, while the corresponding figure for a low content would be5 %.

Neither the TG instrument employed nor the use of a reference curve to correct the experimental curve appears to exert any influence [11].

3.3   References

[1]   N.N.   ISO 11358 (1997)
                                     Plastics – Thermogravimetry (TG) of Polymers –
                                     General Principles

[2]   N.N.   ASTM E 1131 (2003)
                                     Standard Test Method for Compositional Analysis by
                                     Thermogravimetry

[3]   N.N.   DIN 51 006 (2000)
                                     Thermal Anaylsis (TA), Thermogravimetry (TG) –
                                     Principles

[4]   Widmann, G.,   Thermoanalyse
      Riesen, R.   Hüthig Buch Verlag GmbH, Heidelberg 1990

[5]   Hemminger, W.F.,   Methoden der Thermischen Analyse
      Cammenga, H.K.   Springer-Verlag, Berlin, Heidelberg 1989

[6]   N.N.   ASTM E 1582 (2000)
                                     Standard Practice for Calibration of Temperature Scale
                                     for Thermogravimetry

[7]   N.N.   TA Instruments GmbH
                                     Operator's Guide TGA 2950

[8]   N.N.   ASTM D 6370 (1999)
                                     Standard Test Method for Rubber-Compositional
                                     Analysis by Thermogravimetry (TGA)

[9]	Kaisersberger, E., Knappe, S., Möhler, H., Rahner, S.	TA for Polymer Engineering: DSC, TG, DMA, TMA Netzsch Annuals for Polymers No. 3, Selb, Würzburg 1994, p. 36
[10]	Scheirs, J.	Compositionel and Failure Analysis of Polymers John Wiley & Sons, Chichester, England 2000
[11]	Affolter, S., Schmid, M., Wampfler, B.	Ringversuche an polymeren Werkstoffen: Thermoanalytische Verfahren Kautschuk Gummi Kunststoffe 52 (1999) No. 7–8, p. 519

Other References:

[12]	N.N.	ASTM E 1868 (2002) Standard Test Method for Loss-On-Drying by Thermogravimetry
[13]	N.N.	ASTM E 2008 (1999) Standard Test Method for Volatility Rate by Thermogravimetry
[14]	Kopsch, H.	Vergleichende Untersuchungen an Polymeren mit thermoanalytischen Methoden Plaste und Kautschuk 41 (1994), pp. 172–180
[15]	Wunderlich, B.	Thermal Analysis Academic Press, San Diego, 1990
[16]	Pearce, E.M.	Thermal Analysis and Polymer Flammability in Thermal Analysis in Polymer Research and Production
[17]	Utschik, H., Schultze, D., Böhme, K.	Methodes of Thermal Analysis (2): Determination of Mass Transfer CLB Chemie in Labor und Biotechnik, 45, 1994
[18]	Ferriol, M., Gentilhomme, A., Cochez, M., Oget, N., Mieloszynski, J. L.	Thermal Degradation of Poly(Methyl Methacrylate) (PMMA): Modelling of DTG and TG Curve Polymer Degradation an Stability 79 (2003), pp. 271–281

4   Thermomechanical Analysis (TMA)

4.1   Principles of TMA

4.1.1   Introduction

A dilatometer is used to determine the linear thermal expansion of a solid as a function of temperature. Unlike the classical methods in which the experimental setup is kept as free of forces as possible, in thermomechanical analysis, a constant, usually small load acts on the specimen [1]. The measured expansion of the specimen can be used to determine the coefficient of linear thermal expansion α.

The 1st heating phase yields information about the actual state of the specimen, including its thermal and mechanical history. When thermoplastics soften, especially above the glass transition, orientations and stresses may relax, as a result of which postcrystallization and recrystallization processes may occur. On the other hand, the specimen may deform under the applied test load. To determine the coefficient of expansion as a material characteristic, the material must not undergo irreversible changes, such as postcrystallization, postpolymerization, relaxation of orientation or internal stress and so forth during a 2nd heating phase that has followed controlled cooling. In addition, the anisotropy of the molded part must be taken into account; it is advisable to measure in the x-, y- and z-axes. The behavior of plastics on exposure to heat is described in detail in Section 1.1.4.

Coefficient of linear thermal expansion can serve as a material characteristic only where material behavior is reversible.

The coefficient of linear thermal expansion may be recorded as the mean $\overline{\alpha}\,(\Delta T)$ or differential $\alpha(T)$ and is calculated in accordance with DIN 53 752 [2], ISO 113591 Part 1 and 2 [3, 4].

The **mean coefficient of linear thermal expansion** $\overline{\alpha}\,(\Delta T)$ is derived as follows:

$$\overline{\alpha}(\Delta T)=\frac{1}{l_0}\cdot\frac{l_2-l_1}{T_2-T_1}=\frac{1}{l_0}\cdot\frac{\Delta l_{th}}{\Delta T}\quad\left[\frac{\mu m}{m\,^{\circ}C}\right]$$

$\overline{\alpha}\,(\Delta T)$ is an experimental value that varies in each case over the individual temperature measurement range. It is therefore not suitable as a general characteristic or as a general value for calculations but pertains to a specific application.

The temperature-dependent change of length is the progressive change of length expressed in terms of the initial length/reference length l_0. It is a relative measure of the linear expansion, which always has a value of 0 at the start of the trial at the reference temperature T_0. The **differential (or local) coefficient of linear thermal expansion** $\alpha(T)$ is derived as follows:

$$\alpha(T) = \frac{1}{l_0} \cdot \frac{dl_{th}}{dT} \qquad \left[\frac{\mu m}{m\,°C} \right]$$

The values may assume the dimensions $[10^{-6}\,°C^{-1}]$ or $[K^{-1}]$. ISO 11359-2 [4] recommends $[K^{-1}]$, DIN 53 752 [2] $[10^{-4}\,°C^{-1}]$ or $[K^{-1}]$. However, in this chapter, we shall use the unit $[\mu m/(m\,°C)]$ because, in our experience, this conveys a better impression of the magnitudes involved.

Coefficients of linear thermal expansion as a function of temperature are shown in Fig. 4.1 for semicrystalline thermoplastics and in Fig. 4.2 for amorphous thermoplastics. The rise in the coefficient of linear thermal expansion in the glass transition range makes it clear that the values are not constant. For this reason, it is advisable to consider the whole curve when assessing the expansion behavior over a large temperature range.

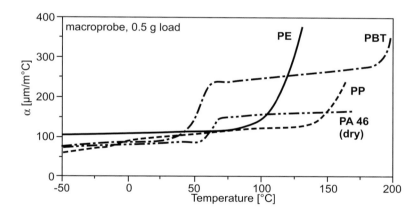

Fig. 4.1   Coefficient of linear thermal expansion for PE, PP, PBT, and PA 46

Macroprobe, specimen crosssection approx. 6 x 6 mm, load 0.5 g,
heating rate 3 °C/min

Characteristics are dependent on temperature range.

The shape of the 1st heating curve is governed by the direction of measurement (see Section 4.2.2.1), the processing conditions, and the thermal and mechanical history (see Section

4.2.3.1). Not only the measurement parameters (see Section 4.2.2.4), but also additives affect the coefficient of linear thermal expansion.

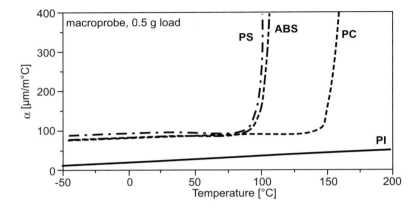

Fig. 4.2   Coefficient of linear thermal expansion of PS, ABS, PC, and PI

Macroprobe, specimen crosssection approx. 6 x 6 mm, load 0.5 g,
heating rate 3 °C/min

Given reversible material behavior, linear expansion is roughly one third of volumetric expansion.

Linear expansion = 1/3 x volumetric expansion

4.1.2 Measuring Principle

A cylindrical or oblong specimen measuring 2–6 mm in diameter or length and usually 2–10 mm in height is subjected to slight loading (0.1–5 g) via a vertically adjustable quartz glass probe. The probe is integrated into an inductive position sensor. The system is heated at a slow rate. If the specimen expands or contracts, it moves the probe. A thermocouple close to the specimen measures the temperature.

Figure 4.3 shows schematic diagrams of the setup for different types of TMA apparatus. The diagram on the left shows the measuring system and the furnace above the specimen. The diagram on the right shows the measuring system located beneath the specimen; for heating, the furnace is lowered on the apparatus from above.

In TMA under load, the measured expansion curve always represents the sum of the different deformation components, such as linear thermal expansion and the load-dependent deformation (force, probe geometry, temperature-dependent modulus).

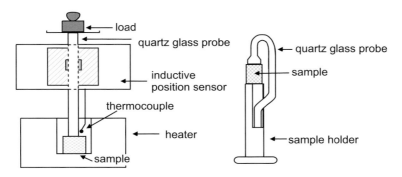

Fig. 4.3   Schematic diagrams of TMA apparatus

   Left: *Apparatus above the specimen*
   Right: *Apparatus beneath the specimen*

Change of linear expansion is the result of linear thermal expansion and load-dependent compression.

Probe shapes differ so as to accommodate various specimen geometries (films, fibers, varying cross-sections) and specific experimental goals (measurement of expansion behavior or of glass transition temperature).

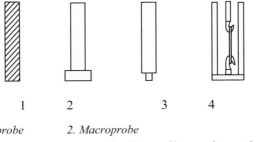

   1   2   3   4

1. Normal probe   2. Macroprobe
3. Penetration probe   4. Setup for films or fibers in the tensile mode

Fig. 4.4   Different shapes of probe used in TMA

The choice of probe depends on the geometry of the specimen. To ensure maximum accuracy when determining the coefficient of linear thermal expansion, it is preferable to use probes with relatively large contact areas. Where the specimen is large enough, a **macroprobe** with a contact area of approx. 28 mm^2 is used. For smaller specimens, for example, with an edge length less than 6 mm, a **normal probe** with a contact area of approx. 5 mm^2 is suitable.

> **Accurate coefficient of linear thermal expansion**
> **Large probe area – low applied load**

Penetration probes with a small contact area of just about 0.8 mm^2 are suitable for determining the glass transition temperature. In the case of amorphous plastics especially, penetration by the probe as the specimen softens yields a signal that is readily evaluated. Owing to the small contact area, even a low applied load generates nonuniform, high compressive stress that does not permit correct sensing and evaluation of the expansion behavior. Therefore the coefficient of linear expansion cannot be determined with accuracy.

Special holders allow thin **films and fibers** to be measured. For this, the specimen is held between two clamps and then suspended in the measuring device. To prevent the thin specimen from twisting or curling, it is loaded with a slight tensile force.

> **Glass transition temperature determination**
> **Linear expansion – large probe contact area – low load**
> **Penetration – small probe contact area – high load**

4.1.3 Procedure and Influential Factors

The stages involved in a TMA measurement are as follows:

- Prepare plane-parallel specimens.
- Measure initial length l_0.
- Load the specimen in the instrument.
- Choose applied load.
- Choose suitable experimental parameters.

The influential factors and sources of error associated with the experiment are explained in detail in Section 4.2.2 with the aid of experimental results from real-life examples.

Factors exerting an influence on the apparatus and specimen are:

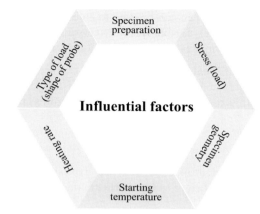

4.1.4 Evaluation

4.1.4.1   Coefficient of Linear Expansion

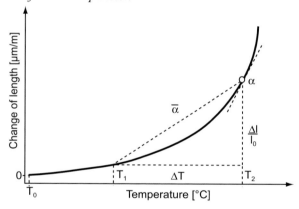

$\alpha(T)$	Differential coefficient of linear thermal expansion:	Tangent at a temperature (T_2), slope at one point
$\overline{\alpha}$ (ΔT)	Mean coefficient of linear thermal expansion:	Slope over a temperature range (ΔT)
$\Delta l/l_0$	Change of length:	Change of length over a temperature range, expressed in terms of the initial length l_0
ΔT	Change of temperature:	Difference between two temperatures (T_2-T_1)

Fig. 4.5   Evaluation of the mean (ΔT) and differential $\alpha(T)$ coefficients of linear thermal expansion

The change of length of the specimen can be used to calculate both the mean coefficient of linear thermal expansion $\overline{\alpha}$ (ΔT) over a specific temperature range and the differential coefficient α (ΔT) of linear thermal expansion (see Section 4.1.1).

Figure 4.5 illustrates the difference between the two characteristics. While $\alpha(T)$ is the respective slope of the tangent at a specific temperature, $\overline{\alpha}$ (ΔT) is the slope of a secant in a temperature range. For this reason, $\alpha(T)$ can be meaningfully plotted as a function of temperature.

> **Mean coefficient of linear thermal expansion: α (ΔT)**
> **Differential coefficient of linear thermal expansion: α (T)**

4.1.4.2   Glass Transition Temperature

At the glass transition, many physical properties of amorphous plastics or of the amorphous domains of semicrystalline thermoplastics undergo stepwise changes in several orders of magnitude. The same applies to their coefficient of linear thermal expansion.

Figure 4.6 shows how loading affects the linear expansion and the differential coefficient of linear expansion of plastics at the glass transition.

Whether the expansion (rising curve) or penetration (falling curve) occur above the glass transition depends on the expansion, probe area, applied load, and the temperature-dependent modulus of elasticity. Generally, unreinforced plastics have a higher coefficient of linear thermal expansion above T_g than below it because of greater segment mobility in the amorphous domains.

Figure 4.6 (left) shows how, in the case of slight loading and a still relatively high modulus of elasticity even above T_g (e.g. in semicrystalline thermoplastics or thermosets), the expansion curve is steeper above T_g. At the same time, there is a step in α-curve.

If the modulus of elasticity changes extensively above T_g, as in the case of thermoplastics and elastomers, the probe can penetrate into the specimen on even slight loading or at least superpose the expansion, Fig. 4.6, right. In such cases, the real expansion behavior can no longer be equivocally measured.

> **In unreinforced plastics, $\alpha(T)$ is**
> **greater above T_g than below it.**

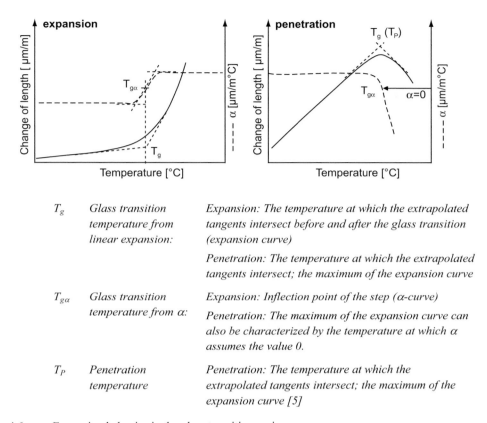

Fig. 4.6   Expansion behavior in the glass transition region

Left: expansion
Right: penetration

The definitions below appear beside Fig. 4.6:

T_g Glass transition temperature from linear expansion:

Expansion: The temperature at which the extrapolated tangents intersect before and after the glass transition (expansion curve)

Penetration: The temperature at which the extrapolated tangents intersect; the maximum of the expansion curve

$T_{g\alpha}$ Glass transition temperature from α:

Expansion: Inflection point of the step (α-curve)

Penetration: The maximum of the expansion curve can also be characterized by the temperature at which α assumes the value 0.

T_P Penetration temperature

Penetration: The temperature at which the extrapolated tangents intersect; the maximum of the expansion curve [5]

Evaluating T_g from the expansion curve

In the literature [4, 6, 7], T_g is evaluated from the expansion curve by determining the point where the extrapolated lines before and after the glass transition intersect. In [8], the intersection of the extrapolated lines on penetration is termed the softening temperature T_s.

Evaluating T_g from the α curve

$T_{g\alpha}$ is calculated from the change of the differential coefficient of linear expansion. For expansion, the turning point of the step is used and, for penetration, the temperature at which $\alpha = 0$ μm/(m °C) is used, as this reflects the maximum of the expansion curve.

4.1.4.3   Test Report

DIN 53 752 contains useful information for compiling a complete test report that describes all experimental parameters and specimen details.

Where appropriate, the test report should contain the following information [2]:
- Reference to the standards employed
- All details necessary for complete identification of the material analyzed
- Shape and dimensions of the specimens, where specimens are made of semifinished and finished products, the position of the specimens in the product and the direction of measurement must be stated.
- Pretreatment of specimens, recording any irreversible changes in length that may have occurred
- Number of specimens tested
- Experimental conditions
- Reference temperature and the pertinent length of the specimen
- Temperature range covered
- Coefficient of thermal expansion $\alpha(T)$ in $[10^{-4}\ °C^{-1}]$ to two decimal places as a function of temperature, on a plot
- Mean coefficient of linear thermal expansion $\overline{\alpha}\ (\Delta T)$ in $[10^{-4}\ °C^{-1}]$ to two decimal places where only one temperature interval was measured
- Diagram of the relative coefficient of linear thermal expansion as a function of temperature
- Details as to whether irreversible changes occurred during the measurements and how many repetitions of the test program were made until linear reversible changes in the length of the specimen occurred
- Date of the test
- Expansion curve

4.1.5 Calibration

The TMA apparatus must be calibrated in terms of length and temperature. If the loading force is applied electromagnetically, it may be calibrated with the aid of known weights. If the apparatus is in constant use, it should be calibrated every month. Because probes have different masses, a new calibration must be performed every time a probe is changed. Calibration and actual test measurements must be performed using the same probe geometry, applied load, heating rate, and temperature range (starting temperature). The specimen should be the same size as the calibration specimen.

Calibration parameters = Experimental parameters

4.1.5.1   Calibrating the Length

The length is calibrated mostly with metallic specimens that have a known, reversible coefficient of linear thermal expansion. Aluminum specimens are frequently used. Their experimental coefficient of expansion in the temperature range of interest is compared with the standard literature value, see also ASTM E 2113 [11]. The calibration factor, F, is calculated as follows:

$$F = \frac{\alpha_{\text{literature}}}{\alpha_{\text{experimental}}}$$

4.1.5.2   Calibrating the Temperature

The temperature is calculated with pure metals (indium, zinc) that have known, reproducible melting points, Table 1.5. The metals should have melting points in the temperature range of interest.

The metal is placed between two sheets (e.g., quartz glass or aluminum) in the apparatus and heated in accordance with the desired experimental parameters. When the melting point of the calibrating metal is reached, the probe compresses the discs, and a pronounced step shows up on the curve. The melting point is the intersection of the tangent to the resultant curve ranges. Several metals can be heated up simultaneously in a sandwich construction for measurement. Temperature calibration is performed with two metals using a penetration probe with an applied weight of 5 g. The temperature is quoted to +/- 1 °C [9], see also ASTM E 1363 [12].

To measure the temperature as close to the specimen as possible, most devices feature a flexible thermocouple. Care must be taken not to hinder expansion of the specimen. After calibration, the position of the thermocouple must not be changed.

4.1.6 Overview of Practical Applications

Table 4.1 shows which characteristics of TMA experiments can be used to assess quality deficiencies and processing flaws or to characterize materials. The scope of TMA is explained in more detail in Section 4.2.3 using real-life examples and the pertinent characteristic curves.

Application	Characteristic	Example
Service temperatures for parts design	α, β	Dimensional stability (constant expansion coefficient)
Expansion behavior in different directions	α	Anisotropy; along and across the direction of processing
Orientation	α	Shrinkage on deorientation $> T_g$

Application	Characteristic	Example
Influence of reinforcing materials	α, β	Change of expansion behavior
Determination of glass transition temperatures	T_g	Change of expansion curve
Material composite	$\alpha_1 \neq \alpha_2$	Adhesive film on support material; coinjection molding, plastic-metal hybrids
Degree of curing in thermosets	T_g	Determination of the glass transition temperature; superposition of shrinkage (through postcuring) and expansion
Physical aging	T_g	Shifting of T_g to higher temperatures owing to reduced void volume in amorphous regions

Table 4.1   Examples of practical applications of TMA in plastics

4.2   Procedure

4.2.1 In a Nutshell

Specimen preparation

The specimen must have plane-parallel contact areas perpendicular to the direction of measurement. The material must not be affected in any way during specimen preparation.

> **Specimen must have plane-parallel contact areas in direction of measurement.**

Specimen geometry

Specimens are usually cylindrical or oblong with a diameter or edge length of 2–6 mm and a height of 2–10 mm. Special equipment is available for measuring films and fibers.

Probe

The coefficient of linear expansion is measured with the aid of normal or macroprobes with a diameter of 2–6 mm. Penetration probes are used for the measuring glass transition temperature, especially of amorphous thermoplastics. Probes and specimen holding equipment are usually made of quartz glass.

Applied load

The load is applied either as a small weight (0.1–5 g) or electromagnetically. The applied load serves either to ensure that the probe makes good contact or to selectively stress the specimen and influences the experimental result.

> **Adapt the probe and applied load to suit the objective and the specimen geometry.**

Heating program

The heating program covers the starting temperature, heating rate, and end temperature. The starting temperature is at least 30 °C lower than the temperature range of interest. The heating rate ranges from 1 to 5 °C max. The end temperature must not be too high, so that the specimen will not be deformed.

Evaluation

The measured parameter is the change of length of the specimen. The mean or differential coefficient of linear expansion is calculated. Transitions such as glass transition temperature, crystallization, and postcuring can be identified.

Interpretation   The 1st heating curve provides information about the thermal and mechanical history of the material. The 2nd heating curve, conducted after controlled cooling, yields material characteristics.

4.2.2 Influential Factors and Possible Errors During Measurement

4.2.2.1   Specimen Preparation

During **sample preparation**, the sample must have plane-parallel contact surfaces in order to guarantee precise determination of the initial length and proper contact by the probe. This is necessary for accurate determination of the initial length and precise contact of the probe. Opposite sides should be parallel to +/- 25 μm [9]. The initial length is measured in the instrument direct at ambient temperature or by the operator using a micrometer. When the specimen is being made from a molded part, care must be taken to avoid thermal and mechanical influences, such as those caused by sawing without cooling or deformation due to clamping in the vise. It is best to employ a water-cooled diamond saw or mortising machine to remove the specimen carefully. Any flash should be removed with a scalpel. The desired direction of measurement should be known before the specimen is taken.

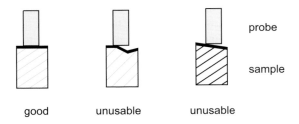

Fig. 4.7   Contact surfaces for TMA measurement

Prepare specimen carefully;
plane-parallel contact surfaces in the direction of measurement.

Figure 4.8 shows the influence of the direction of measurement on the expansion curves for a tensile bar of ABS. Plane-parallel test specimens were worked out from the solid along the x-, y-, and z-axes. The mean coefficient of linear expansion in the range 0–100 °C varies with the direction of measurement from 61 to 80 µm/(m °C). When the glass transition temperature is exceeded, the specimens expand extensively along the y- and z-axes, but shrink along the x-axis. The reason is that molecules along this axis become highly oriented during processing and relax at elevated temperatures.

The reference length (initial length) l_0 of the specimen is measured at ambient temperature. The change of length is 0. Prior to the experiment, the specimen is cooled to a starting temperature of 20 °C. On cooling, the specimen contracts, giving rise to a negative linear expansion at the starting temperature.

To prevent confusion between shrinkage and penetration (penetration by the probe: see Fig. 4.6), the applied load should be as low as possible (see Section 4.2.2.4).

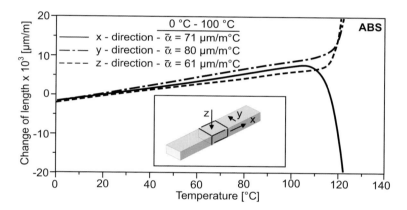

Fig. 4.8   Anisotropic expansion of an injection molded ABS tensile bar

Cooling to starting temperature, 1st heating, normal probe, l_0 = 4.02 to 4.12 mm, specimen crosssection 4 x 4 mm, applied load 2 g, heating rate 3 °C/min

Anisotropic specimens: Measure in several directions.

If TMA studies are to be used for gaining information about the effect of relaxation and shrinkage processes in a sample, careful thought needs to be given prior to sample preparation to the point at which sampling will occur. Most molded parts exhibit different expansion behavior as a function not only of processing direction but also of the cross-section. For example, when information is needed about prior processing, it may be sensible to consider expansion behavior of the edge region relative to the center of a molded part, Fig. 4.9. The faster-cooled boundary layer shrinks much more than the middle layer.

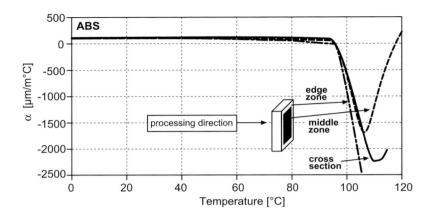

Fig. 4.9   Difference in expansion of a molded ABS tensile rod in the edge and middle zones

Cooling to starting temperature, 1ˢᵗ heating scan, normal probe, applied load 0.5 g,
heating rate 2 °C/min

To escape the influence of the processing direction, measurements are often made in the thickness or z-direction. However, Fig. 4.9 shows that, even with measurements in the thickness direction, the side on which the probe makes contact with the sample must be chosen carefully. In one direction, the sample, which was originally flat, curves so much that a "bridging" effect occurs. The sample arches further and further away from the sample holder. This gives the impression of expansion that actually is not present.

In the case of soft, thin samples, a heavier load may have to be applied, but it must be borne in mind that this might affect the coefficient of expansion (Section 4.2.2.4).

4.2.2.2   Specimen Geometry

The dimensions of a TMA specimen are primarily dependent on the geometry of the molded part. The choice of probe to use can be made only after the possible contact surface of the specimen has been established.

Specimens are usually cylindrical or oblong (diameter or edge length of 2–6 mm) and have a height of 2–10 mm. They should be large enough to permit good resolution – a large diameter gives the specimen mechanical stability. Bulky specimens, however, increase the temperature gradient within the specimens during heating or cooling. Specimen dimensions should be the same for comparative measurements.

Figure 4.10 shows the measurements for two PC samples of different height, but otherwise subjected to the same experimental conditions. There is a huge difference in expansion. Even at around 40 °C, relaxation can be detected in a very small sample. This leads to superimposition of expansion and relaxation, which in turn leads to much lower expansion coefficients. Extensive relaxation is registered in both cases just before the glass-transition temperature is reached. In the thicker sample, this effect is shifted about 10 °C to a higher temperature, probably due to slower warming of the sample. This effect is not due to penetration of the sample by the probe, as confirmed by inspection of the sample surface.

The purely material-based expansion coefficients are measured in the 2nd heating scan after defined cooling (see Section 4.2.3.6).

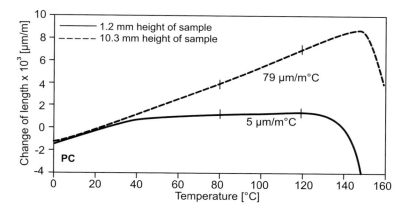

Fig. 4.10   Influence of geometry on the curve of change in length of PC samples of different height

Normal probe, specimen crosssection approx. 5 x 4 mm, heating rate 3 °C/min, 1st heating scan

The geometry required for measuring film and fibers is dictated by the intended clamping devices. Not only is uniform tightening of the jaws important (sequence, tightening torque), but also proper sample removal, as stretched fibers and film frequently exhibit highly directional expansion behavior (see also Section 4.2.3.6).

4.2.2.3   Probe

A range of probe shapes are available in TMA to suit different specimen geometries and problems (see Fig. 4.4). Probes with a relatively large contact area (2–6 mm in diameter) make for highly accurate determination of the coefficient of linear expansion. Penetration probes with small contact area (approx. 1 mm in diameter) are employed in the measurement of softening temperatures, particularly of amorphous thermoplastics. This is because the probe can penetrate into the specimen when T_g is reached.

Several other clamping devices enable fibers and films to be measured. However, as the mounted specimens are usually very small, the major influence of the clamping effects restrict high measurement accuracy.

Choose probe to suit specimen geometry and test objective.

4.2.2.4   Applied Load

An appropriate load may be applied to the probe, which is located on the center of the specimen. The weight of the probe is counterbalanced by a spring. Depending on the apparatus employed, the load is applied either by manually adding weights or electronically by a solenoid. The load maintains direct contact between probe and specimen as the specimen expands or contracts. This requires the application of a low applied force at the very least; ASTM E 831-2003 [9] recommends 1–3 g. The force counteracts positive expansion of the specimen body and reinforces contraction, especially above the glass transition when the elastic modulus decreases sharply. To ensure exact determination of the coefficient of linear expansion, virtually no force must work against expansion.

ISO 11359-3 [5] recommends 50 g for determination of penetration temperature.

Applied force: Ensures direct contact between probe and specimen

Figure 4.11 shows how the applied force affects the change of length of a semicrystalline thermoplastic. The specimens, whose crosssectional area had an edge length of approx. 2 mm and so roughly matched the area of the normal probe, were loaded with different weights (0.5 g, 10 g, 50 g). The elastic modulus curve shown was obtained by means of DMA and represents the change of elastic modulus in the T_g region.

At different loading below the glass transition, that is, up to roughly 40 °C, the curves are found to have the same slopes and thus the same expansion coefficients. This also applies to temperatures above T_g up to roughly 150 °C.

Although the specimens are compressed by the applied load in relation to their height, the expansion coefficient is virtually the same in all three cases. This is because the elastic modulus in this rubber-elastic temperature and condition range changes only slightly. Despite progressive softening of the intervening amorphous ranges, the crystallites ensure strong bonding at a comparatively high elastic modulus level.

Applied load [g]	α (ΔT = -20 °C–30 °C) [μm/(m °C)]	α (ΔT = 90 °C–150 °C) [μm/(m °C)]
0.5	118	357
10	118	358
50	118	331

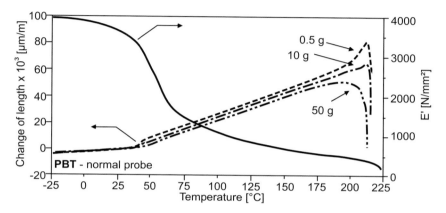

Fig. 4.11   Influence of the applied load on the temperature-dependent change of length of PBT
               above and below T_g

               Cooling to starting temperature, 1st heating phase, normal probe, l_0 = 4.00 to 4.18 mm,
               Specimen crosssection approx. 2 x 2 mm, heating rate 3 °C/min

Semicrystalline thermoplastics

**Below T_g: Applied load exerts hardly any influence. Above T_g: Applied load has
little influence initially, but this increases with the start of the melting range (T_{im}).**

Similar measurements were performed on amorphous ABS (Fig. 4.10). Here, too, the slope
in the change of length curve as far as the glass transition is comparable at all loading levels.

Above T_g, the higher applied load (0.5, 5, and 10 g) causes the curves to drop whereas at
low loading the specimen can expand even further. The reason for this is extensive softening
in the glass transition range, as also revealed by the dynamic mechanical analysis (DMA)
curve; the elastic modulus falls to almost zero. The measurements show that, for accurate
information about the coefficient of linear expansion or the expansion behavior of plastics,
the applied load should be as low as possible.

Applied load [g]	α (ΔT = 0 °C–70 °C) [μm/(m °C)]	α (ΔT = 120 °C–130 °C) [μm/(m °C)]
0.5	102	1790
10	99	1548
50	100	1285

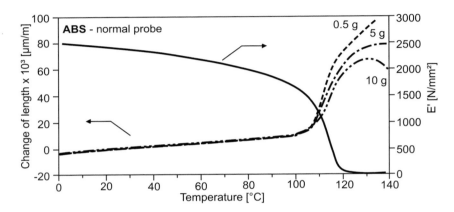

Fig. 4.12   Influence of the applied load on the temperature-dependent change of length of ABS, above and below T_g

Cooling to starting temperature, 1st heating phase, normal probe, l_0 = 4.22 to 4.26 mm, specimen crosssection approx. 2 x 2 mm, heating rate 3 °C/min

---

Amorphous thermoplastics

Below T_g: Applied load exerts low influence.
Above T_g: Applied load exerts large influence.

A small applied load is also better when determining the T_g of semicrystalline thermoplastics, too, as the change of expansion is hampered less. This is shown in Fig. 4.11 for PBT. The jump in the α-curve becomes much clearer.

Experience also shows that other effects attributable to thermal history, such as annealing, degradation of internal stress, and relaxation, can be represented more clearly by a low applied load.

Thermoset materials undergo a much lower change in elastic modulus and creep behavior above the glass transition than do thermoplastics. The applied load therefore has less influence. Nevertheless, it should be as low as possible when their thermal history is being characterized.

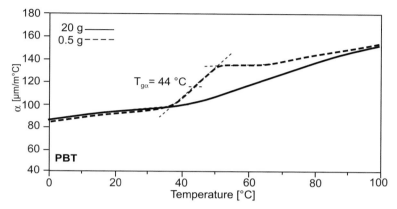

Fig. 4.13   Influence of an applied load on the change of the coefficient of linear expansion $\alpha(T)$ in the glass transition range of PBT

Macro probe, $l_0 = 4.18$ mm, specimen crosssection approx. 6 x 6 mm, heating rate 3 °C/min

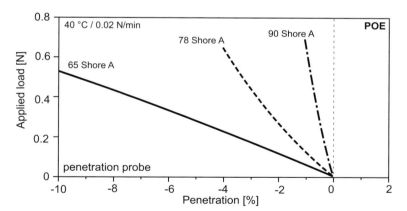

Fig. 4.14   Force-dependent dimensional change during increasing-load trials on POE of different Shore hardness values

Penetrating probe, $l_0 = 3.5$ mm, specimen crosssection approx. 4 x 4 mm, measuring temperature 40 °C, load increase 0.02 N/min

In addition to a constant, specified load, it is possible to conduct trials involving increasing loads. These could, for example, be performed as small tensile tests on films or with standard or penetrating probes on rectangular samples. Figure 4.14 shows the force-related change in dimensions of three POE samples of different Shore hardness.

The experiments were performed isothermally at 40 °C, and the load was increased at a rate of 0.02 N/min. A penetrating probe that sinks into the sample was used, which is why the dimensional change has a negative algebraic sign.

Some TMA instruments offer the option of dynamic load changing. Such load modulation provides insights into viscoelastic effects in the sample.

4.2.2.5   Measuring Program

The **starting temperature** must be at least 30 °C lower than the measuring range of interest. It should be maintained until no further changes in specimen expansion are measured. The time for this varies from 10 to 20 minutes according to the temperature.

The **heating rate** should not exceed 5 °C/min on account of the relatively large specimen [9]. If a faster heating rate is necessary, the specimen must have a smaller crosssection.

When comparative measurements are being performed, the heating programs must be identical because the parameters of the measurement influence the result. Figure 4.12 illustrates the influence of the heating rate on PMMA. Although temperature and length calibrations were performed for each heating rate, differences due to heating rate were measured in the middle coefficient of thermal expansion.

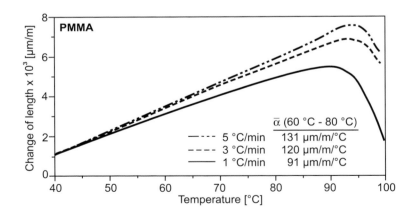

Fig. 4.15   Change of length of PMMA as a function of heating rate

Macro probe, $l_0 = 6.37$ to 6.80 mm, specimen crosssection approx. 6 x 6 mm, applied load 2 g

The **end temperature** should be only as high as absolutely necessary because otherwise the specimen geometry will be influenced by deformation and the experimental result will be inaccurate. Any evaporating gases may contaminate the sensitive measuring assembly of the TMA (if located above the specimen).

The 1st heating phase provides information about the thermal and mechanical history, processing influences, service conditions, and so forth. After controlled cooling at a constant cooling rate, the 2nd heating phase reveals material-specific characteristics (at least as far as the end temperature of the 1st heating phase).

As a result of the speciemen´s thermal history or if the end temperature is too high, the specimen geometry may change after the 1st heating phase. If this happens, a fresh specimen may have to be prepared.

> **1st heating phase: Thermal and mechanical history**
> **2nd heating phase: Material-specific characteristics (after controlled cooling))**

In the case of critical specimens and at high temperatures, it is best to conduct the measurement in an inert gas to prevent oxidation, see [4]. Similarly, at very low starting temperatures, helium may be used as a **purge gas** because of its better cooling action.

4.2.2.6   Evaluation

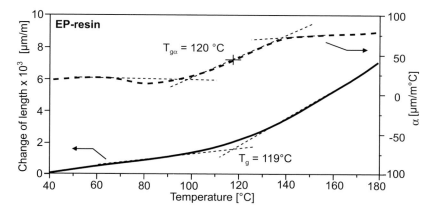

Fig. 4.16 T_g determination of EP resin from the curves for the change of length and the coefficient of linear expansion $\alpha(T)$

Macro probe, $l_0 = 5.61$ mm, specimen crosssection approx. 6 x 6 mm, applied load 0.5 g, heating rate 3 °C/min

In Section 4.1.4.1, we discussed the evaluation of the differential and mean coefficient of linear thermal expansion as set out in [2] and [9]. There are two established methods for determining the glass transition temperature (see Section 4.1.4.2): evaluation of the change of length (TMA curve) or of the derived coefficient of linear expansion $\alpha(T)$. These are now presented again with the aid of concrete examples, Fig. 4.16.

The change of length curve is preferred for the evaluation as it involves the direct measurement signal. The intersection point of the tangents yields a glass transition temperature of $T_g = 119$ °C; from the α-curve, $T_{g\alpha} = 120$ °C.

Figure 4.17 illustrates how the T_g is determined from penetration of PVC. After initial expansion, the probe penetrates into the specimen at the T_g.

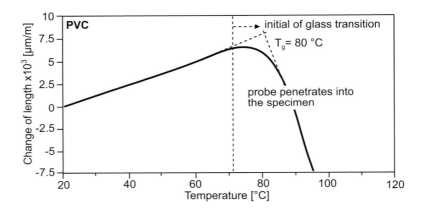

Fig. 4.17 T_g determination of PVC from the change of length curve

Penetration probe, $l_0 = 4.12$ mm, specimen crosssection approx. 4 x 4 mm, applied load 5 g, heating rate 5 °C/min

4.2.3 Real-Life Examples

4.2.3.1   *Thermal and Mechanical History*

As in DSC, the 1st heating phase in a TMA measurement provides information about the thermal and mechanical history of the specimen. A 2nd heating phase, preceded by controlled cooling, yields information about purely material behavior, as shown in Fig. 4.18 for PBT.

In the 1st heating curve, the change of the coefficient of linear expansion in the glass transition range is influenced by superposed effects, for example, relaxation, cold crystallization, and so forth.

After controlled cooling at a constant cooling rate, the 2nd heating curve measures material-specific behavior with an undisturbed glass transition.

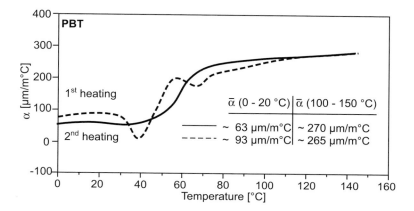

Fig. 4.18   1st and 2nd heating of PBT to reveal history

Normal probe, specimen crosssection approx. 4 x 4 mm, applied load 0.1 g, heating rate 3 °C/min

Figure 4.19 is a further example of the influence of thermal and mechanical history, this time on a PMMA specimen.

In the 1st heating phase, the specimen begins to shrink on reaching the glass transition at 100 °C. From 120 °C on, the change of length curve continues with a greater slope. This effect is not detectable either during cooling or in the 2nd heating phase. As usual, the curve has a slight step in the glass transition range.

Note: Measurement of cooling curves becomes problematic if the specimen deforms at the end of the 1st heating phase on account of its history and/or elevated tempera-ture, such that the specimen's shape is different from the original geometry and cannot be corrected at high temperatures. Although many instruments can heat at constant rates, they cannot cool uniformly − even though this is just as critical.

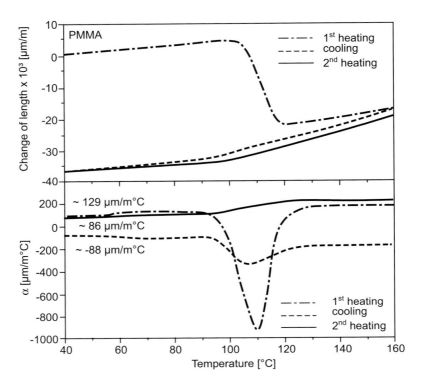

Fig. 4.19   Change of length and the coefficient of linear expansion of a PMMA specimen during
controlled 1st and 2nd heating and cooling

*Macro probe, l_0 = 8.34 mm, specimen crosssection approx. 6 x 6 mm, applied load 2 g,
heating or cooling rate 3 °C/min*

The upper part of Fig. 4.20 demonstrates the influence of crystallization processes that occur
during a TMA measurement. A PPS specimen (non-postcured) cooled rapidly from the melt
begins to expand again on being reheated but when the T_g is passed at 70 °C it shrinks due
to postcrystallization. When crystallization is complete, the specimen continues to expand
and the coefficient of linear expansion rises. The lower part of Fig. 4.20 shows the TMA
curve for annealed PPS. Crystallization had been triggered by a separate annealing process
in the furnace, and the purpose of the following TMA measurement on the annealed
specimen was to determine the glass transition temperature and the expansion behavior. No
superposing due to postcrystallization processes occurs. However, the crystalline domains
produced during annealing at 200 °C begin to melt just above this temperature, at 210–215
°C (see also Section 1.2.3.3). This is identifiable from the brief increase in the coefficient of
linear expansion caused by the increase in volume.

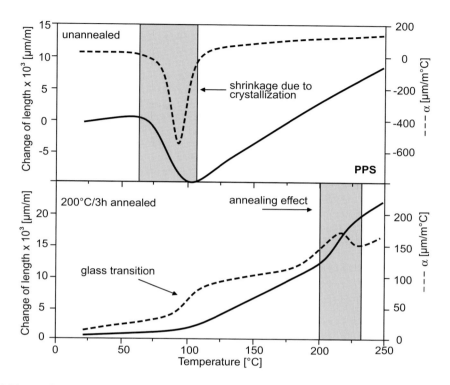

Fig. 4.20   Change of length and differential coefficient of linear expansion of amorphous PPS and
              PPS annealed for 3 h at 200 °C

              Macro probe, l₀ =3.04 mm, specimen crosssection approx. 6 x 6 mm, applied load 2 g,
              heating rate 3 °C/min

Postcrystallization leads to shrinkage because the volume changes.

Because of their different molecular arrangements and density, annealed and not annealed
specimens have different coefficients of linear expansion. This is illustrated in Fig. 4.21 for
PET; in both cases, the degree of crystallization was determined by means of DSC (see
Section 1.2.3.2). The mainly amorphous specimen has a crystallinity of 6% and a coefficient
of linear expansion of 72 μm/(m °C) whereas the annealed specimen has a crystallinity of
24% and a coefficient of linear expansion of 49 μm/(m °C).

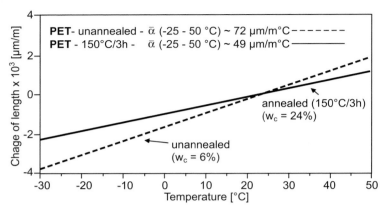

Fig. 4.21   Influence of crystallinity on the expansion behavior below T_g for PET, annealed
            for 3 h at 150 °C, and untreated

*w_c = Crystallinity from DSC measurement, expressed in terms of ΔH_m^0 = 140 J/g,
Cooling to starting temperature, 1st heating phase, macro probe, l_0 = 1.51 to 1.98 mm,
specimen crosssection approx. 6 x 6 mm, applied load 0.5 g, heating rate 3 °C/min*

4.2.3.2   Influence of Conditioning

Figure 4.22 shows the influence of water content on the experimentally determined coefficients of linear expansion of polyamide 46 specimens subjected to different conditioning.

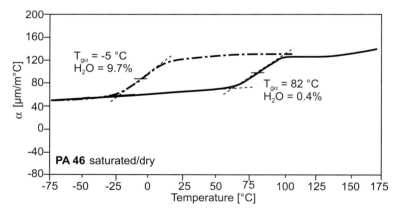

Fig. 4.22   Differential coefficient of linear expansion α(T), PA 46 conditioned saturate and dry

*Cooling to starting temperature, 1st heating phase, macro probe, specimen crosssection
approx. 6 x 6 mm, applied load 5 g, heating rate 5 °C/min*

The differential coefficient of linear expansion of dry PA 46 containing 0.4% water undergoes a steplike rise from roughly 75–125 μm/(m °C). As read off from the α-curve, the glass transition temperature is $T_{g\alpha} = 82$ °C.

The $T_{g\alpha}$ of the saturat specimen (9.7% H_2O) is −5 °C and the coefficient of linear expansion also increases from roughly 75–125 μm/(m °C). At temperatures of approx. 90 °C and above, the bound water starts escaping from the polyamide, a process that may cause the specimen to deform (blistering may occur) and thus lead to nonreproducible α values.

Note: *Diffusion of water from the inside of the specimen to the outside and evaporation from the surface becomes increasingly retarded and so are not uniform.*

4.2.3.3   Influence of Aging

Aged PA 66 specimens were to be used to examine the influence of aging and selective sampling. The specimens were conditioned for 16 weeks at 120 °C, which produced marked discoloration on their surfaces. These discolored boundary layers were measured in one case (whole crosssection) and sanded off in another to ensure that only the central, nondiscolored specimen zone (without boundary layer) was measured.

Fig. 4.23 shows the change in the coefficient of expansion for these two specimens in the thickness direction (z-direction). Aside from the glass transition range, incipient crystallization induced by the aging process can be seen at approx. 120 °C.

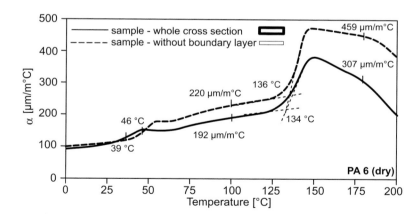

Fig. 4.23   Differential coefficient of linear expansion α(T) of aged PA 66 specimens (16 weeks at 120 °C); measurement in the z-direction; "whole crosssection" (with boundary layer) and "without boundary layer" (center)

1st heating scan, macroprobe, specimen crosssection approx. 6 x 6 mm, load 5 g, heating rate 5 °C/min

The coefficients of expansion are the same below T_g, whereas, above it, the "whole cross-section" specimen has lower values, a fact that may be attributable to greater specimen crystallinity (boundary layer undergoes pronounced postcrystallization and thereby increases the crystallinity of the whole specimen). The glass transition temperature $T_{g\alpha}$ is shifted to a value roughly 7 °C lower in the "whole crosssection" specimen. Provided the conditioning states are identical, such a shift indicates a degradation effect. Because the degradation occurs more in the boundary layer, $T_{g\alpha}$ of the "whole crosssection" specimen occurs at a lower temperature.

4.2.3.4   Curing of Thermosets, Postcuring

TMA provides important information about the state of curing of thermosets. In addition to determining the glass transition temperature as a measure of the degree of curing, we can evaluate the coefficient of linear expansion at temperatures below T_g to draw conclusions to be drawn about the state of curing, Fig. 4.24.

Here, the specimens were each cured for 2 h at different temperatures (23, 60, 100, and 120 °C). In the subsequent TMA studies, the glass transition temperature was obtained from the expansion curve by the method of intersecting tangents and the mean coefficient of linear thermal expansion was obtained from a temperature range below the T_g.

As in DSC and DMA, the glass transition temperature increases with rise in curing temperature, while the coefficient of linear expansion decreases.

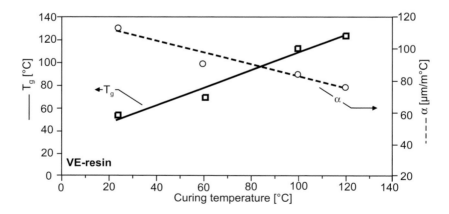

Fig. 4.24   Characterizing the state of curing of VE resin from the T_g and α ($T < T_g$)

Normal probe, specimen crosssection approx. 4 x 4 mm, applied load 5 g, heating rate 5 °C/min

> **Higher degree of curing**
>
> **Smaller coefficient of linear expansion below T$_g$**
> **T$_g$ shifts to higher temperatures**

Incompletely cured thermosets frequently show shrinkage-induced postcuring effects during the 1st heating phase.

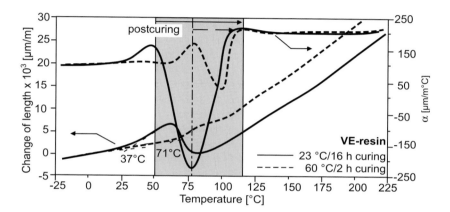

Fig. 4.25   Influence of postcuring on the change of length and coefficient of linear expansion of incompletely cured VE resin

Cooling to starting temperature, 1st heating phase, normal probe, $l_0 = 7.00–7.10$ mm, specimen crosssection approx. 4 x 4 mm, applied load 5 g, heating rate 5 °C/min

As shown by the two examples of differently cured VE resins in Fig. 4.25, expansion in the range of the respective current glass transition overlaps with the decrease in length due to postcuring.

The specimen cured at ambient temperature exhibits extensive shrinkage on reaching the glass transition temperature of 37 °C. This effect occurs at much higher temperatures and is less pronounced in the sample cured at 60 °C. Its glass transition temperature is 71 °C.

4.2.3.5   Influence of Fillers and Reinforcing Materials

It is difficult to measure fiber-reinforced specimens because of their anisotropy and the inhomogeneity of the specimens. While glass transition temperatures can be evaluated fairly reliably, this is frequently not the case for the coefficient of linear expansion. This is

illustrated in Fig. 4.26 for a glass-fiber-reinforced PA GF30, which was analyzed perpen-
dicularly and parallel to the direction of reinforcement.

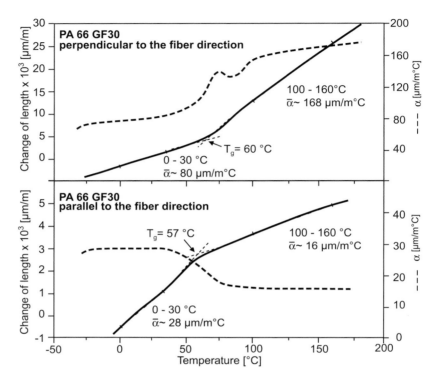

Fig. 4.26   Linear expansion and coefficient of linear expansion of PA 66 GF30, perpendicular and
               parallel to the fiber direction (injection direction)

               Cooling to starting temperature, 1ˢᵗ heating phase, normal probe, l_0 = 9.20–10.07 mm,
               specimen crosssection approx. 4 x 4 mm, applied load 5 g, heating rate 5 °C/min

Perpendicular to the direction reinforcement, the usual curve for a semicrystalline PA is
obtained, with a change in the linear expansion at the glass transition (Fig. 4.26, top). The
coefficient of linear expansion above and below the T_g at ~168 and ~80 μm/(m °C) are
comparatively high for reinforced material and more or less reflect the shape of unreinforced
PA. In other words, the polymer matrix dominates.

By contrast, parallel to the direction of reinforcement, the linear expansion curve has an
unusual shape. Below the T_g, the specimen has an expected coefficient of linear expansion
of ~ 28 μm/(m °C) that becomes even smaller, however, when the glass transition

temperature is exceeded. This behavior is often observed in reinforced or filled plastics and is due to the interaction of the reinforcing compounds and the polymer matrix adhering to them. Above the T_g, the elastic modulus of the matrix is much lower and is less able to contribute to expansion or is more readily restrained during expansion by the stiffer fibers aligned in the same direction. Nevertheless, the glass transition temperature here can be evaluated by the tangent intersection method.

Aside from filler orientation, the shape of the fillers and fibers exerts a major influence on the expansion of the specimen, Fig. 4.27.

As previously described, glass fibers cause a substantial change in the expansion properties of plastics. Above the glass transition zone, they lead to a reduction in the coefficient of expansion despite greater mobility and expansion of the matrix.

Glass beads and mineral fillers are relatively uniform in shape and exert a similar reinforcing action in all directions, but this is not as pronounced as in the case of glass fibers. It is striking that greater relaxation is evident in the case of materials reinforced with the glass beads and mineral substances. This may be explained by the uneven temperature distribution of the fillers that conduct heat better. The decrease in the coefficient of expansion from about 70 °C indicates incipient postcrystallization of the PBT.

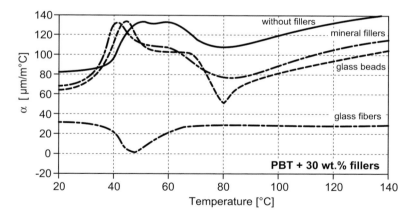

Fig. 4.27   Coefficient of linear expansion of PBT with 30 wt.% glass fibers, glass beads and mineral fillers, measured in the injection direction

Cooling to starting temperature, 1st heating, normal probe, l_0 = 6 mm, specimen crosssection approx. 4 x 10 mm, applied load 2 g, heating rate 3 °C/min

The quantitative influence of metallic fillers on the coefficient of expansion is shown in Fig. 4.28 for various blends of sPS and SrFe. As expected, the coefficient drops as the filler content rises, but there is also a shift in the glass transition temperature $T_{g\alpha}$ from about 108 °C to about 88 °C. The start of the glass transition falls from around 87 °C to 82 °C, a fact that may be attributed to the greater thermal conductivity of the metallic filler, or could be due to chain degradation caused by greater friction during processing.

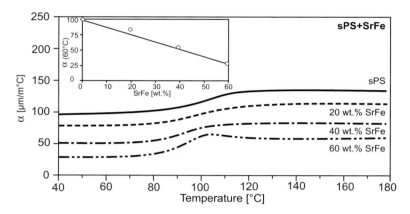

Fig. 4.28   Temperature-dependent coefficient of expansion of sPS and blends with SrFe

2nd heating, normal probe, l_0 = 6 mm, specimen crosssection approx. 4 x 10 mm, applied load 2 g, heating rate 3 °C/min

4.2.3.6   Expansion and Shrinkage of Fibers and Film

Fibers and film for certain applications are made by deliberately freezing in shrinkage forces or a degree of orientation. One such application is shrink-wrap film. The molded part for packaging is wrapped in the film, which is then immobilized and heated. The film initially expands as the temperature rises, but then shrinks as molecular mobility is dissipated [10].

TMA can be used to investigate these processes thermally, as a function of direction and, to a certain extent, as a function of force.

Figure 4.29 illustrates this effect as a function of direction for the example of a shrink-wrap tube. Measurement of the specimen along its length revealed barely any change in length over the temperature range shown. At around 80 °C, brief expansion indicative of a postcuring effect may be seen. The tube was probably stretched at temperatures between 70 and 80 °C. As measured in the circumferential direction, shrinking appears to have started at about 70 °C and is measurable as a negative change of length. From about 90 °C, the material expands further in both directions of measurement.

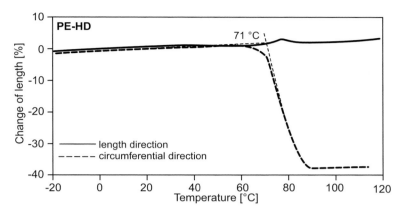

Fig. 4.29   Influence of the direction of measurement on the expansion and shrinkage of
a shrink-wrap tube of PE-HD

Tensile tester, l_0=8.4 mm, specimen crosssection approx. 0.75 x 6 mm,
heating rate 3 °C/min, 1st heating scan

4.2.3.7   Investigating Deformation

When an external force acts on plastics to deform them, it has three distinct, overlapping
components:

  —   Spontaneous elastic deformation
  —   Time-dependent viscoelastic or relaxation deformation
  —   Time-dependent viscous deformation

The deformation phenomena are characterized by the molecular deformation and damage
mechanisms that occur in the polymer. Pure elastic deformation stems from spontaneous
changes in the distances between atoms and from valence angle distortion of the strong
chemical bonds. In time-dependent viscoelastic or relaxation deformation, the molecules or
molecular groups need a certain amount of time to undergo molecular rearrangement to
produce the deformation corresponding to the acting stresses [10].

This means that there is a certain time lag before plastics respond to an imposed stress.
Relaxation occurs when delayed attainment of equilibrium causes a measurable parameter to
diminish. Because the load may be a stress or a strain, the relaxation process is referred to as
stress or strain relaxation. The length of time needed for this process is called the relaxation
time. When a body is deformed, a relatively high amount of stress builds up inside that
gradually relaxes as time passes (stress relaxation). In strain relaxation, imposed stresses
lead to a time-dependent deformation known also as retardation or creep. Additive superimposition of individual deformation components is permissible only to a certain load level.

The limit up to which the processes may be correctly described both mathematically and physically is the limit of the linear-viscoelastic range [10].

TMA can determine loading and relaxation deformation in a very low loading range at definite temperatures. The following diagram illustrates this for PE-HD and PE-LD film, both of which were loaded for 30 min under a tensile force of 0.5 N, before the load was removed. From the change in length, it is possible to record how the specimen responded to this loading and unloading. The evaluation consists in calculating the percentage change in length at maximum load and at the end of unloading.

While such trials cannot replace genuine creep tests, they can provide preliminary information about different materials or material modifications. Because the measurements are similar and the specimens are small, measurements can be performed fairly quickly at different temperatures.

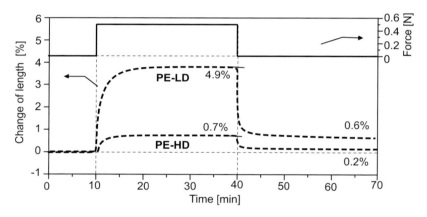

Fig. 4.30   Creep studies on PE-HD and PE-LD film

Tensile tester, specimen crosssection 2.20 mm², applied load 0.5 N, loading time 30 min, unloading time 30 min, temperature 40 °C

4.3   References

[1] Hemminger, W.F,   Methoden der Thermischen Analyse
     Cammenga, H.K.   LABO 4, 1990

[2] N.N.   DIN 53 752 (1980)
                                  Testing of Plastics – Determination of the Coefficient of
                                  Linear Thermal Expansion

[3]   N.N.   ISO 11359-1 (1999)
                                  Plastics – Thermomechanical Analysis (TMA) –
                                  Part 1: General Principles

[4]   N.N.   ISO 11359-2 (1999)
                                  Plastics – Thermomechanical Analysis (TMA) –
                                  Part 2: Determination of Coefficient of Linear Thermal
                                  Expansion and Glass Transition Temperature

[5]   N.N.   ISO 11359-3 (2002)
                                  Plastics – Thermomechanical Analysis (TMA) –
                                  Part 3: Determination of Penetration Temperature

[6]   Wunderlich, B.   Thermal Analysis
                                  Academic Press, Inc., San Diego 1990

[7]   Turi, E.   Thermal Characterization of Polymeric Materials,
                                  Second Edition
                                  Academic Press, Inc., Orlando 1997

[8]   Seyler, R.J.   Assignment of the Glass Transition
                                  ASTM, Philadelphia 1994

[9]   N.N.   ASTM E 831 (2003)
                                  Standard Test Method for Linear Thermal Expansion of
                                  Solid Materials by Thermomechanical Analysis

[10] Stellbrink, K.K.U.   Micromechanics of Composites
                                  Composite Properties of Fibre and Matrix Constituents
                                  Carl Hanser Publishers, Munich 1996

[11] N.N.   ASTM E 2113 (2002)
                                  Standard Test Method for Length Change Calibration of
                                  Thermomechanical Analyzers

[12] N.N.   ASTM E 1363 (2003)
                                  Standard Test Method for Temperature Calibration of
                                  Thermomechanical Analyzers

Other References:

[13] N.N.   ASTM E 1545 (2000)
                                  Standard Test Method for Assignment of the Glass
                                  Transition Temperature by Thermomechanical Analysis

[14]   N.N.   ASTM E 1824 (2002)
                                   Standard Test Method for Assignment of a Glass
                                   Transition Temperature Using Thermomechanical
                                   Analysis Under Tension

[15]   Oberbach, K.,   Saechtling: Kunststofftaschenbuch, 29th Edition
       Baur, E.,   Carl Hanser Publishers, Munich 2004
       Brinkmann, S.,
       Schmachtenberg, E.

[16]   Ehrenstein, G.W.   Mit Kunststoffen konstruieren, Second Edition
                                   Carl Hanser Publishers, Munich 2001

[17]   Ehrenstein, G.W.,   Duroplaste
       Bittmann, E.   Aushärtung, Prüfung, Eigenschaften
                                   Carl Hanser Publishers, Munich 1997

[18]   Riga, A.T.,   Materials Characterization by Thermomechanical
       Neag, C.M.   Analysis
                                   ASTM, Philadelphia, PA, 1991

[19]   Price, D. M.   Modulated-temperature – Thermomechanical Analysis
                                   Thermochimica Acta 357/358 (2000), pp. 23–29

5   pvT (pressure-volume-Temperature) Measurements

5.1   Principles of pvT Measurements

5.1.1   Introduction

The major polymer-processing techniques subject the materials to different temperatures T (mostly heating followed by cooling), large fluctuations in pressure p that can exceed 1000 bar, and, during processing, phase transformations from solid to molten and back to solid again. An important quantitative variable is the mass density ρ or specific volume $v = 1/\rho$, which expresses the dimensions of the molten polymer in the mold or of the frozen polymer in the finished part in terms of its mass. The changes that occur in density with change in processing parameters therefore hold the key to the dimensional accuracy of the finished parts.

The acting pressure p, the ambient temperature T, and the resultant specific volume v are sufficient to characterize the polymers and are commonly plotted in an eponymous pvT diagram, with the specific volume mostly plotted against the temperature while the pressure serves as a parameter, Fig. 5.1 and Fig. 5.2.

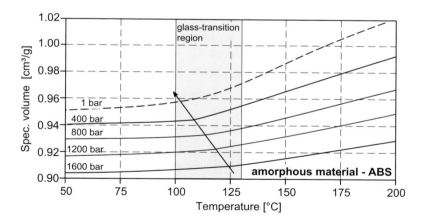

Fig. 5.1   pvT measurement of an amorphous ABS as a function of temperature and pressure

Cooling rate 2 °C/min, pressure 400–1600 bar,
the 1-bar curve was obtained by extrapolation

pvT: Determination of specific volume as a function of temperature and pressure

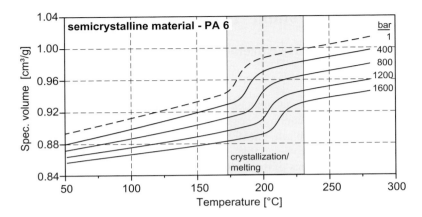

Fig. 5.2   pvT measurement of a semicrystalline PA 6 as a function of temperature and pressure

Cooling rate 2 °C/min, pressure 400–1600 bar

Transformations: Glass transition and melting

Volume of polymers

The volume of a polymer has three components:

- the volume of the macromolecular chains,
- the vibrational volume generated by thermal vibration of the molecules,
- the free volume arising from defects and voids between the macromolecules.

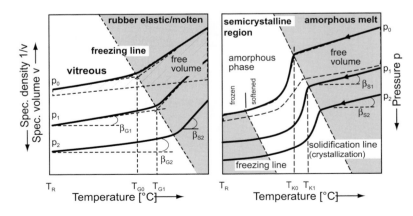

Fig. 5.3   Typical pvT diagram of an amorphous (left) and a semicrystalline (right) polymer,
                isobaric cooling and heating

As the melt cools, the volume of the macromolecular chains does not change, but a change does occur in vibrational volume, which varies with the temperature, and diminishes until absolute zero is reached. The free volume declines down to a certain limiting value and then remains almost constant. At this limit, called the glass transition, the voids are so small that molecular segments cannot undergo rearrangement. Consequently, the segments and also the voids and defects more or less freeze by themselves.

Cooling at temperatures above the glass transition is temperature-dependent, and shrinking occurs due to decreases in the free volume and the vibrational volume of the macromolecules. Below the glass transition, shrinking is due only to the reduction in vibrational volume. The volume continues to decrease almost linearly, but the rate of decrease is much reduced, Fig. 5.3.

This decrease in volume on cooling corresponds to the coefficient of thermal expansion. It is measured by means of TMA in one-dimensional directions in solids (Chapter 4).

Cooling rate

Amorphous polymers

The temperature at which the glass transition occurs during cooling varies with the rate of cooling and with the acting pressure. The influence of the cooling rate is discussed below. During the cooling process, attainment of thermodynamic equilibrium is hindered by diminishing vibrational energy and chain mobility, especially in the neighborhood of the glass transition. The more closely the glass transition is approached, the longer the relaxation time of the macromolecules. Faster cooling equates to a shorter measuring time and thus a higher measuring frequency. The glass transition is a dynamic process, that is, it depends on the observation frequency and thus, in our case, on the cooling rate. It shifts to higher temperatures when measurements are made at higher frequencies, that is, at faster cooling rates. During rapid cooling, the polymer is somewhat farther from the equilibrium state than it is during slow cooling.

Semicrystalline polymers

If the structure of the polymer allows the macromolecular chains to align themselves in uniform, 3 dimensional array, the polymer may crystallize on cooling from the melt. This process is time dependent as it is first necessary for the crystallization nuclei to be formed from which the crystals are created by further chain attachment. The rates of nucleation and crystal growth depend on the temperature. Because nucleation is necessary, polymer crystallization occurs below the melting temperature; this is called supercooling. The crystallization temperature, too, depends on the cooling rate. A higher cooling rate shifts the crystallization temperature to lower temperatures. In the crystallization process, the heat of crystallization generated has to be dissipated. The crystals formed during crystallization have a higher density and thus a lower specific volume than the amorphous zones. This means that the specific volume of semicrystalline polymers decreases during crystallization. In line with the usual crystallization rates, the associated change in volume therefore occurs more rapidly at first during crystallization (cooling); conversely, it is slower during melting (see also Chapter 1 DSC).

Pressure

If pressure is additionally acting on the melt, the free volume and the vibrational amplitude of the macromolecules during cooling are reduced to a certain degree. The critical free volume, that is, the point at which molecular segment mobility is hindered, is reached at higher temperatures. Consequently, the glass transition and crystallization are shifted further to higher temperatures. The free volume, along with the vibrational amplitude of the macro-molecules, is reduced as a function of the acting pressure, both above and below the glass transition. The outcome of this is that the slope of the lines of constant pressure becomes smaller or flatter with increasing pressure, Fig. 5.3.

5.1.2 Measuring Principle

pvT readings may be obtained in two ways:

 – Compression of the specimen in a cylinder by a piston – **direct method**, Fig. 5.4, left,

 – High-pressure dilatometry with a confining liquid (usually mercury) as a pressure transfer medium – **indirect method**, Fig. 5.4, right

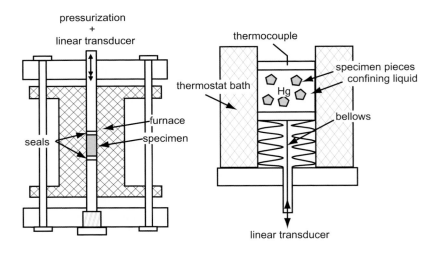

Fig. 5.4   Schematic diagrams of pvT measurements

         Left: *direct method*
         Right: *indirect method*

Direct Method

The specimen is located in a heatable cylinder, embedded between two seals. It is com-pressed by a piston moving in the direction of the cylinder axis. The bottom of the cylinder is securely sealed. Expansion and contraction of the specimen under applied temperature and pressure cause deflections in the piston that are measured by an attached linear transducer and act as a measure of the change in specific volume. The deflections and the specimen mass are used to calculate the absolute specific volume of the specimen.

$$\Delta v \ (p,T) = \Delta l \ (p,T) * \pi * r^2 \ / \ m$$

The direct method assumes that the pressure across the specimen does not drop. This is rather unlikely to happen with solids and needs to be borne in mind when the readings are being evaluated.

Before it is introduced into the cylinder, the specimen usually has to be melted to prevent voids from forming in the cylinder and errors from being generated when the conversion is carried out to the change in volume. The direct method for determining changes in volume as a function of pressure and temperature therefore generally requires elimination of thermal history.

Indirect Method

The analyte material in the chamber is surrounded by a confining liquid (usually mercury) that transmits the pressure isotropically to the specimen. The bottom of the chamber comprises a bellows.

The pressure is transmitted into the chamber by means of oil and the bellows. A furnace surrounding the chamber provides the necessary equilibration. The change in volume of confining liquid and specimen is transmitted via the bellows to a linear transducer and measured inductively.

Calibrating the device with the chamber completely full, for example, with mercury, enables volume changes that are not due to the specimen to be canceled arithmetically during data processing. The measured parameters are the change in volume $\Delta v(p, T)$ between the reading $v(p_{exp.}, T_{exp})$ and the specific volume at the start $v \ (p_{ini}, T_{ini})$ when the device is set up. The calculation proceeds via the change in specific volume between the reading (p_{exp}, T_{exp}) and the starting conditions (p_{ini}, T_{ini}).

$$\Delta v \ (p, T) = v \ (p_{exp.}, T_{exp.}) - v \ (p_{ini.}, T_{ini.})$$

In order that the absolute specific volume may be determined from the measured volume changes, the density of the analyte material must be determined in a separate experiment at room temperature and normal pressure [1].

The specimen occupies a volume of approx. 1 cm^3 in the pvT chamber and may be investigated in almost any form (granules, fragments, liquid). However, care must be taken to ensure that the material is free of occluded air and is not too finely divided (powder) [1]. It is possible to perform measurements on specimens in the form delivered, that is, without prior melting.

5.1.3 Procedure and Influential Factors

The steps involved in conducting a direct pvT measurement are:

- prepare the specimen (e.g., gentle granulation),
- determine the amount to weigh out,
- precompress the seals,
- introduce the specimen into the device,
- preequilibrate and precompress the specimen,
- choose a suitable measuring program,
- weigh the specimen after measurement.

Factors influencing instruments and specimens are:

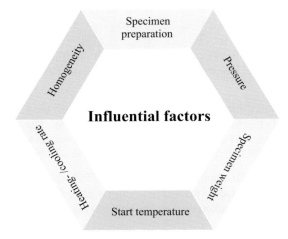

Influential factors and sources of experimental error are discussed in detail with the aid of curves from actual experiments in Section 5.1.6.

5.1.4 Evaluation

5.1.4.1   General Parameters Obtained with pvT Diagrams

The basic physical behavior of amorphous and semicrystalline plastics during a pvT experiment is shown schematically in Fig. 5.3.

Because pressure is always acting on the specimen during the measurement, the volume curve for a pressure of 1 bar has to be calculated. This is generally done by extrapolation. It may lead to errors in transition areas that exhibit a pressure-dependent temperature shift. In such cases, the curves have to be specially adjusted.

pvT diagrams are required for calculating the following physical parameters.

Density ρ:

$$\rho(p, T) = \frac{1}{v(p, T)}$$

Coefficient of volumetric expansion β:

$$\beta = \frac{1}{v}\left(\frac{\Delta v}{\Delta T}\right)_p$$

Isothermal compressibility ξ:

$$\xi = -\frac{1}{v}\left(\frac{\Delta v}{\Delta p}\right)_T$$

5.1.4.2   *Test Report*

The test report should contain the following information:

- All information necessary for complete identification of the material analyzed
- Pretreatment of the material before the test
- Mass of the specimen before the measurement
- Melting temperature
- Precompression pressure and time
- Sealant material
- Rate of temperature change (programmed heating or cooling before and during the measurement) or test temperature for isothermal method
- Type of measuring program (isothermal, isobaric)
- Substance used for temperature calibration
- Volume, pressure, and temperature change, experimental curve
- Any special observations with regard to device, test conditions, or specimen behaviour
- Mass of specimen after measurement
- Date of the test

5.1.5   Calibration

The volume, pressure, and temperature of pvT devices need to be calibrated. Calibration must be performed under the same conditions used for the real test, that is the heating or cooling rate, specimen volume, and sealant material must be the same.

5.1.5.1   *Volume Calibration*

Exact determination of the volume before and during a pvT measurement requires a precise knowledge of the geometry of the capillaries (this applies only to the direct method). These should therefore be checked at regular intervals with calibrated linear measuring devices so that the calculation program may be corrected as necessary.

Calibration substances are available for the linear transducer.

Before every measurement, the volume of the sealant (direct method) or the confining liquid (indirect method) must be determined and corrected for the pressures and temperatures obtaining during the measurement. It must be ensured that PTFE seals, especially, are pre-compressed in order that irreversible effects may be ruled out.

5.1.5.2   Temperature Calibration

Standard calibration substances are also used for the thermocouples. With the direct method, it is possible to measure the temperature inside the material by means of a thermocouple attached directly to the piston, which ensures precise temperature correction, Fig. 5.5.

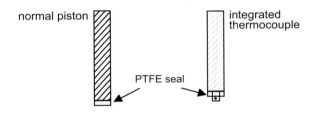

Fig. 5.5   Normal piston and temperature-measuring piston for direct pvT

5.1.6   Overview of Practical Applications

Application	Characteristic	Example
Knowledge of the pressure dependence of the specific volume for estimating shrinkage during processing	T_g, T_{ic}, β, $\rho(t,T)$	Adjusting the machine parameters of an injection molding machine
Determining model coefficients for simulation programs for allowing for the melt compressibility and calculating shrinkage		Empirical TAIT method, IKV method, FOV theory
Determination of shrinkage of thermosets as a function of pressure and temperature	T_{ic}, β, $\rho(t,T)$	Influence of inhibitors on the shrinkage of thermosets

Table 5.1   Examples of practical applications of pvT measurements on polymers

5.2   Procedure

The following section deals primarily with the direct method, as the authors have gained practical experience of this type of instrument.

5.2.1   In a Nutshell

Specimen preparation
Granules, powder, or liquids may be measured. Molded parts must be granulated gently. The specimens must be packed tightly into the capillary (direct method) with as few voids as possible.

Specimen mass
The specimen mass is determined from the optimum height (volume) and the density of the material. The calculated mass must be weighed out exactly and should be kept constant for comparative measurements. For thermoplastics, the specimen mass is approx. 8 g. With reactive systems, the chemical reaction is heavily dependent on the weight and so it is vital that comparable amounts are used. At the end of the measurement, the specimen has to be reweighed.

> **Keep specimen masses comparable; reweigh specimen at end of experiment.**

Seals
Seals prevent the material from escaping. They are generally made of PTFE. Polyimide (PI) seals are used for temperatures above 290 °C.

The system and the seals must be precompressed before the start of the experiment.

Measuring program
In addition to a starting and end temperature, a cooling/heating rate or pressure/pressure range have to be chosen.

Pressure-dependent cooling is usually employed for thermoplastics (cooling at different pressures).

Reactive systems are usually measured isothermally, or with heating without applied pressure, or at a specified pressure.

Starting temperature
Thermoplastics (cooling mode): A temperature 20–30 °C above the melting temperature is chosen; the specimen must be molten.

Reactive systems (heating mode): These are usually started at room temperature.

Heating rate/cooling rate   To ensure uniform equilibration of the specimen, cooling/ heating rates of 1–5 °C/min max.

> **Heating/cooling rate are of 5°C/min max.**

End temperature   The end temperature depends on the type of experiment. In heating mode, the experiment should be terminated before the start of decomposition. In cooling mode, room tempera- ture may be the lowest point; after thermoplastics have crystallized and the material is solid, there is no point measuring beyond around 50 °C.

Pressure   Pressures ranging from 200 to 2500 bar are feasible. The chosen pressure depends on the objective. Thermoplastics are typically determined with cooling curves constructed from five different pressures.

The specimen is precompressed before the measurement, usually at a pressure of around 200 bar.

> **Pressures ranging from 200 to 2500 bar**

Evaluation/ Interpretation   The pvT curves are usually obtained at different pressures. Their shape is interpreted and they are extrapolated to a pressure of 1 bar. This 1-bar curve provides information and values for simulation.

5.2.2   Influential Factors and Possible Errors During Measurement

5.2.2.1   Specimen Preparation

Granules, powder, or liquids may be measured. These have to be carefully packed into the capillary with as few voids as possible (direct method). Molded parts up to a certain size (approx. 1 cm^3) may be measured by the indirect method. If the direct method is used, the molded parts have to be granulated. Granulation should be as gentle as possible.

Polymers that absorb water must be dried beforehand to prevent evaporation during the experiment. It is advisable to follow drying procedures for processing or to measure rheological properties, such as MVR.

If solid specimens have to be measured by the direct method without loss of thermal history, the diameters of the specimens and the capillaries should be as identical as possible. Fuchs [2] recommends filling the residual interstices with mercury. The expansion behavior of this confining liquid must then be determined separately.

5.2.2.2   Specimen Mass

The specimen mass has a major influence on pvT measurements. Aside from the change in temperature distribution within the specimen due to the cooling influence of the piston (direct method), another key variable is the loss of pressure along the specimen length, especially in the direct method.

As for curing reactions of thermoset systems, it must be remembered intrinsic specimen heating varies closely with the specimen mass and influences the result of the measurement. This means that specimen masses should be as similar as possible when reactive systems are being compared.

5.2.2.3   Seals

PTFE seals are generally used for temperatures ranging from room temperature to 290 °C. PI seals are used for higher temperatures. For direct measurements, the seals are pre-compressed in the capillary prior to each measurement and are corrected by the initial length of the specimen. Where measurements are based on a thermocouple introduced into the melt (direct method), the upper seal must be perforated so that the thermocouple can be positioned directly in the melt.

5.2.2.4   Measuring Program

In pvT studies, it is usual to start above the melting or glass transition temperature and to cool. An isothermal dwell time is needed to allow the temperature equilibrium to become accurately established. The dwell time ensures that the specimen is completely melted and uniformly equilibrated, but not thermally damaged.

After the specimen has been melted or preequilibrated, it is precompressed to allow any trapped air bubbles to escape. Again, both the duration and the level of the applied pressure must suit the material to prevent it from being damaged before the measurement. A 1-minute precompression at a pressure of approx. 500 bar is usually sufficient.

Like the heating rates, the cooling rates in pvT experiments must be kept relatively low (1– 5 °C/min max.) to ensure homogeneous temperature distribution within the specimen. Where shrinkage is a prime consideration, it should be remembered that crystallization is heavily influenced by the cooling rate.

At high cooling rates, the temperature region of high nuclei growth occurs is passed quickly. This leads to extensive athermal nucleation that in turn leads to a large number of nuclei. The spherolites formed from the crystallites impede each other's growth and the result is a particularly fine spherolitic microstructure. This impairment of crystallization lowers the degree of crystallization and increases the specific volume. The effect diminishes with rise in pressure. DSC measurements performed after the pvT measurements confirmed this relationship for the example shown in Table 5.2.

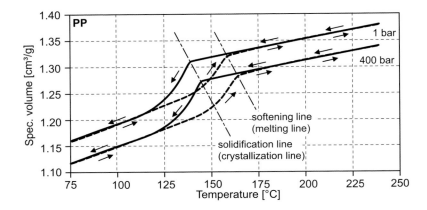

Fig. 5.6   Isobaric cooling (←) and heating measurements (→) of PP in a direct pvT measurement

Heating and cooling rate 3 °C/min

1-bar curve extrapolated from 400-bar and 800-bar curves

Cooling rate/ Pressure	1 °C/min	5 °C/min	10 °C/min
200 bar	78 J/g	74 J/g	72 J/g
500 bar	76 J/g	75 J/g	74 J/g
1500 bar	77 J/g	77 J/g	76 J/g

Table 5.2   Heat of fusion of PA 6 samples with different thermal histories (cooling rate, pressure) measured on specimens weighing approx. 3 mg in a DSC instrument under a heating rate of 10 °C/min

The "crystallization state" produced during rapid cooling attempts to relax to the specific volume of the optimally infinitely small cooling rate as a function of time and temperature. This leads to postcrystallization and to undesirable postshrinking in practice [3].

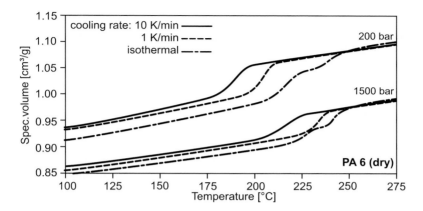

Fig. 5.7   Change in specific volume of PA 6 as a function of cooling rate

pvT experiments frequently involve several cycles of different pressures during a single measurement. Aside from the considerable amount of time involved, this places severe thermal stress on the specimen, which is why the melting temperature is so important. Figure 5.8 illustrates this problem for the example of POM-H at a melting temperature of 220 °C. Thermal degradation of POM releases formaldehyde, which evaporates. After the first four measurements at 500 bar, which proceeded normally, there is an apparent increase in volume until the pressure of the evaporating formaldehyde becomes so great that it can escape through the seals.

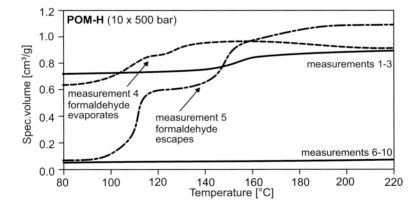

Fig. 5.8   10 repeat measurements on POM-H, maximum temperature 220 °C

            Cooling rate 5 °C, pressure 500 bar

When the measurements are performed at a maximum temperature of 190 °C, this effect is not even observed after the 10th cycle. When the initial temperatures are relatively low, care must again be taken to ensure that the material can completely melt. It may be advisable to briefly heat the specimen at elevated temperatures before the measurement and then to restrict the measuring cycle to a somewhat lower temperature.

5.2.2.5   Evaluation

pvT measurements are evaluated by examining the curves recorded at different pressures. The shape of the cooling curves is particularly interesting for injection molding as it can serve to estimate the shrinkage during crystallization and allows relevant machine parameters to be selected. Conditions at ambient pressure would be of interest for simulation programs. In that case, it is necessary to extrapolate various pressure curves to 1 bar. The pvT data are used to provide the coefficients for modeling the thermal expansion behavior.

5.2.2.6   Comparison with TMA

TMA and pvT experiments generally have different scopes. Whereas TMA is used for studying the directional dependency of expansion behavior, thermal history, and so on, pvT examines volume changes after the thermal history has been eliminated. The experimental setup also has something to do with this. TMA requires "dimensionally stable" specimens, such as samples cut from molded parts, whereas pvT specimens have to have a certain geometry because of the apparatus used. In pvT, the specimen is usually melted before the measurement to generate the right geometry (for direct pvT) or to avoid occluded air.

> **TMA: Determination of directionally dependent expansion behavior of solids.**
> **pvT: Determination of volume change preferably in the melt state or during crystallization.**

Sometimes, we are interested in examining expansion or shrinkage of a material across various physical states. Figure 5.9 compares pvT and TMA studies performed on polycarbonate (PC).

For the comparison, the TMA was performed on PC in three directions. The TMA measurements pertained to the second heating phase, following defined cooling, to eliminate the thermal history as much as possible. The 1-bar curve was extrapolated so that the pvT values could be compared.

Application of pressure (especially below the T_g) in the direct method proved problematic because of pressure loss in the specimen. Extrapolation to 1 bar was therefore calculated from the experimental curves for 400 and 200 bar (standard) and from the curves for 1600 and 800 bar.

The TMA and pvT curves do show agreement in the coefficient of volumetric expansion over the range extending from about 60–140 °C. From about 145 °C, the specimen becomes so soft because the glass transition temperature has been reached. The TMA instrument is no longer able to determine expansion because the specimen starts to flow or the piston penetrates into the specimen. The pvT curves show much greater fluctuation, especially in the temperature range below T_g, that is, in the solid range. This is probably where most of the problems with occluded air and application of the pressure occur. Extrapolation to 1 bar from the curves for 1600 bar and 800 bar show smaller fluctuations among the readings, as well as a shift in the glass transition temperature to higher temperatures on account of the higher applied pressure.

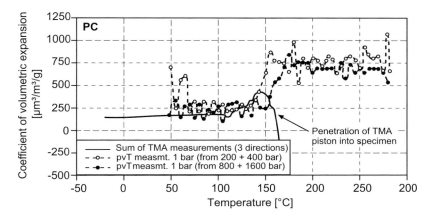

Fig. 5.9   Comparison of the specific volume of PC as determined by pvT and TMA

(TMA: sum of expansion coefficients in three directions)
Heating and cooling rate 3 K/min, applied load (TMA) 0.5 g, normal piston

The coefficient of volumetric expansion (ß) may be calculated from TMA readings as the sum of the coefficients of linear expansion in three directions (x, y, z):

$$\beta = \alpha(x) + \alpha(y) + \alpha(z)$$

5.2.3   Real-Life Examples

5.2.3.1   Shrinkage During Cooling of Polymers

The following overview illustrates the value and the change in specific volume of various semicrystalline thermoplastics, each of which was measured in a favorable temperature range. The curves are 1-bar curves extrapolated from 1000-bar and 2000-bar curves. The sudden leap in specific volume in the region of crystallization is particularly evident.

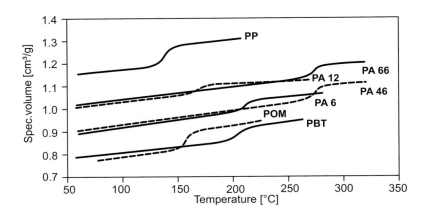

Fig. 5.10   Change in specific volume of different semicrystalline plastics during cooling

Cooling rate 3 °C/min, 1-bar curves extrapolated from 1000-bar and 2000-bar curves, direct method

Figure 5.11 provides a similar overview for amorphous thermoplastics. Freezing of these polymers shows up as a change in the slope of the curve.

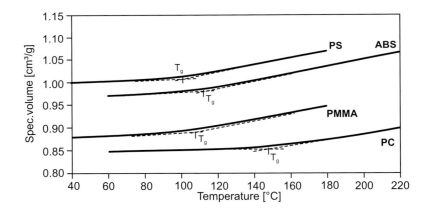

Fig. 5.11   Change in specific volume of various amorphous polymers during cooling

Cooling rate 3 °C/min, 1-bar curves extrapolated from 1000-bar and 2000-bar curves, direct method

5.2.3.2   Shrinkage During Curing of UP Resin

Thermoset materials shrink during curing, which is an exothermic process. This shrinkage during the chemical reaction is pressure dependent and may be measured by means of pvT. The heating mode is used in this case, in contrast to the cooling mode usually employed for thermoplastics.

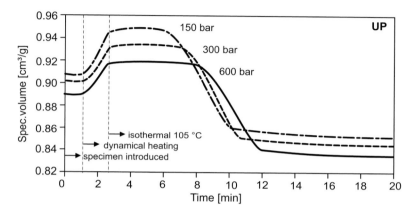

Fig. 5.12   Specific volume for pressurized dynamic curing of UP resin

Heating rate 5 °C/min, specimen mass approx. 0.8 g
direct method, heating rate DSC 20 °C/min

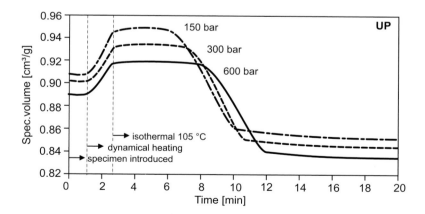

Fig. 5.13   Specific volume for pressurized isothermal curing of UP resin

Isothermal 105 °C, heating rate 40 °C/min, specimen mass approx. 0.8 g, direct method

The curing reaction of the system described in Fig. 5.12 takes place over a temperature range of 110–140 °C; the recorded DSC curve has been included in the diagram to illustrate the relationships. It is difficult in this type of experimental setup (dynamic heating) to discern a dependence of the start of reaction (onset temperature) on the applied pressure; the effect on the specific volume and thus the density is clear right up to the cured state.

Conducting measurements under isothermal conditions would make it easier to identify the start of the reaction. Isothermal pvT measurements are more sensitive at revealing the pressure-dependence of the start of the reaction.

The UP resin under discussion shows a shift in onset temperature to longer times as the pressure increases, Fig. 5.13.

5.2.3.3   Influence of Fillers on Shrinkage

Because the density of fillers and reinforcing materials is different from that of the matrix, they have an influence on the specific volume.

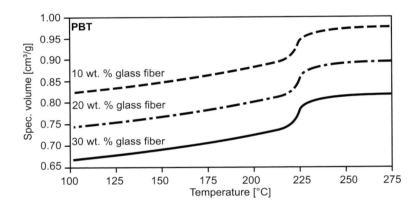

Fig. 5.14   Isobaric cooling of PBT with different glass-fiber contents

Cooling rate 3 °C/min, 1-bar curve extrapolated from 800-bar and 1600-bar curves, direct method

Figure 5.14 shows this influence for the case of PBT with different glass-fiber contents. The whole curves are shifted parallel to each other as a function of glass-fiber content.

Not only the filler content, but also the nature and shape of the filler affect the shape of the pvT curve, Fig. 5.15.

Ultimately, in this case, it is the volume of the filler that is responsible for the different experimental curves. Furthermore, the influence of the thermal conductivity of the fillers on the start of crystallization is discernible. The poorer conducting mineral filler causes a shift in crystallization to somewhat lower temperatures.

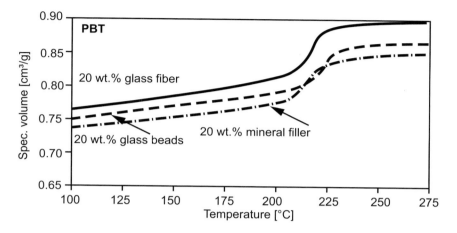

Fig. 5.15   Isobaric cooling of PBT with different filler shapes

Cooling rate 3 °C/min, 1-bar curve extrapolated from 800-bar and 1600-bar curves, direct method.

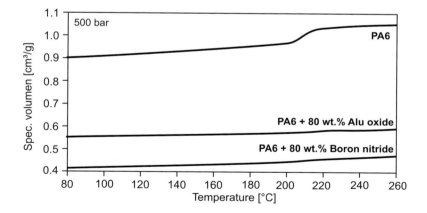

Fig. 5.16   Isobaric cooling of PA 6 with different fillers

Cooling rate 3 °C/min, 500 bar, direct method

The extent to which shrinkage during crystallization can be influenced by different types of fillers is illustrated by the pvT curves (Fig. 5.16) for pure PA 6 matrix material and very highly filled blends with aluminum oxide and boron nitride (80 wt. %). The large difference in the absolute specific volume is due to the high density of the two fillers.

Shrinkage behavior of polymer blends

Figure 5.17 shows pvT curves for each of PA 66 and POM (50/50) and a 50/50 blend of the two.

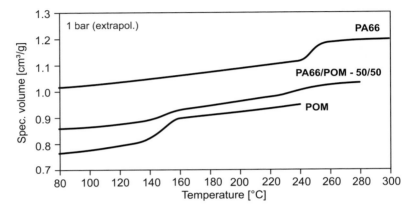

Fig. 5.17   Isobaric cooling of PA 66, POM and a 50/50 blend of PA 66/POM

Cooling rate 3 °C/min, 1-bar curve extrapolated from 500-bar and 1000-bar curves, direct method.

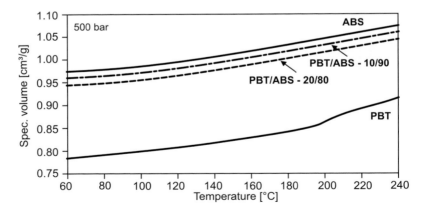

Fig. 5.18   Isobaric cooling of PBT/ABS blends

Cooling rate 3 °C/min, 500 bar, direct method

The pure materials show the typical jump in the curve for the crystallization region. The blend retains both crystallization regions, except that the height of the jump is reduced in line with the relative proportions. The curve for the specific volume lies between the curves for the two pure components.

Blends of ABS containing 10 and 20% of PBT yield the curves shown in Fig. 5.18. The amorphous ABS dominates the curves for the blends. The step in the crystallization of PBT is hardly noticeable at low blending ratios.

5.2.3.4   Change in State During Injection Molding

pvT diagrams can be used to illustrate the change in thermodynamic state during the three injection molding processes of mold filling, holding phase, and residual cooling phase [2].

This is accomplished by making a pvT plot of the pressures and temperatures that occur in the various subprocesses.

Figure 5.19 shows the change in pressure measured near the gate and the calculated mean temperature change in the molded part during the injection molding cycle. The labeled points are transferred to the pvT diagram for the material, Fig. 5.20.

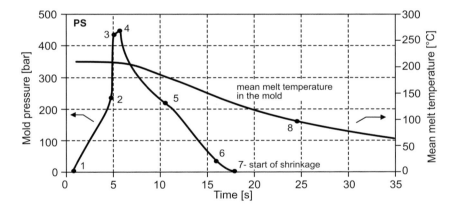

Fig. 5.19   Pressure and temperature change in the mold during mold filling and the cooling phase for injection molding of PS [10]

Curves of this kind showing the change in state of each subprocess yield important information about changes in shrinkage and melt mobility.

Point 7 (attainment of atmospheric pressure) shown in the diagrams is important where the start of shrinking is concerned. It may occur at different temperatures, the temperature depending on the holding pressure profiles or cooling conditions. Shrinkage can thus be influenced accordingly.

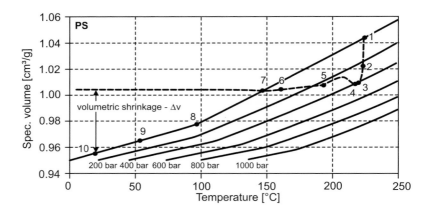

Fig. 5.20   Change in state during the injection molding process, after Stitz [8, 9].

1   Start of mold filling (melt makes contact with pressure sensor)
2   Cavity is volumetrically filled
3   End of compression phase
4   Switchover to holding pressure (partial discharge of mold, melt flow into melt antechamber)
5   Holding pressure level
6   Gate is frozen
7   Pressure fallen to atmospheric pressure (start of shrinking)
8   Average melt temperature has reached freezing temperature
9   Demolding
10   Molded part has reached room temperature

5.2.3.5   Temperature Increase During Adiabatic Melt Compression

pvT may be used to calculate temperature rises during adiabatic compression [4].

If a sudden pressure increase (Δp) occurs in a melt, and if heat exchange between melt and the environment is ignored, adiabatic compression is said to have occurred. The following equation may be used to approximate the temperature rise (ΔT) caused by compression.

$$\Delta T = \frac{\alpha T}{\rho \; c_p} \cdot \Delta p$$

In this equation, whose derivation will not be explained, c_p is the specific heat at constant pressure and T is the absolute temperature in Kelvin. The temperature increase comprises both the reversible and the irreversible (i.e., heat generated by internal friction) components of the pressure rise.

The equation for the irreversible component ΔT_{irr} is given by the Voigt–Kelvin model, as set out in [5], as follows:

$$\Delta T_{irr} = \frac{\xi}{\rho\, c_p} \Delta p^2$$

This returns temperature increases for most polymer melts (irreversible and reversible) of 5–20 °C for sudden compression from 1 to 1000 bar.

For polystyrene having the following material data:

$$\alpha \approx 5.5 \times 10^{-4} \quad K^{-1}$$
$$\rho \approx 1.01 \times 10^{3} \quad kg/m^3$$
$$c_p \approx 1.95 \quad kJ/kg\ grd$$
$$\xi \approx 6.3 \times 10^{-4} \quad mm^2/N$$

the following temperature increase is obtained at a temperature of 227 °C and a pressure increase to 1000 bar in line with the equation above (constant material values are assumed):

$$\Delta T = 14.0\ °C.$$

The irreversible component of the sudden pressure increase to 1000 bar, according to the above equation, is:

$$\Delta T_{irr} = 3.2\ °C.$$

This value is a very good match for temperature increases measured in the melt in the screw antechamber as a result of sudden compression during injection mold filling [6].

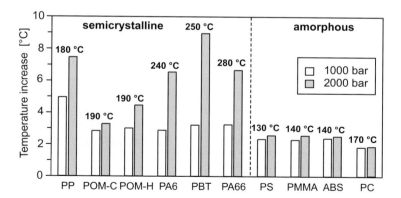

Fig. 5.21   Adiabatic temperature increase in the melt for pressurization of different polymers.

Isothermally approx. 20 °C above the respective melting temperature and glass transition temperature, pressurization 1000 bar, 2000 bar, direct method

Trials of this kind were replicated in the pvT instrument. For this, semicrystalline and amorphous thermoplastics were kept isothermally at roughly 20 °C above the melting temperature or glass transition temperature and pressurized at 1000 or 2000 bar. A thermo-couple located in the piston registered the temperature rise during pressurization. The results for semicrystalline and amorphous thermoplastics are summarized in Fig. 5.21.

Temperature increases of up to 15 °C were measured. The higher the chosen pressure, the greater was the temperature increase. The temperature rise for amorphous thermoplastics occurs at much lower values and proves to be less dependent on the pressure in comparison to the corresponding figures for the semicrystalline thermoplastics.

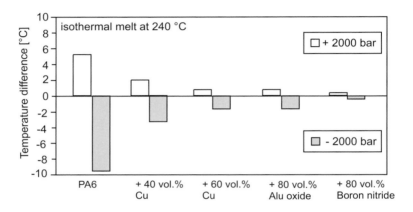

Fig. 5.22   Temperature rise in the melt in the case of pressurization and pressure release
                 of 2000 bar on PA 6 containing fillers of differing thermal conductivity

                 Isothermal 240 °C, pressurization +2000 bar and release –2000 bar, direct method

Aside from the temperature increase on pressurization, it is interesting to determine how the temperature in the melt behaves when the pressure is released. If the melt is expanded spontaneously, an immediate drop in melt temperature occurs, that is, supercooling. With PA 6, a temperature rise of roughly 5 °C occurs on pressurization at 2000 bar, and of around 10 °C on expansion from 2000 bar to 1 bar. This rise and drop in temperature was measured for PA 6 and blends of PA 6 containing different types of fillers that have a pronounced effect on thermal conduction, and with different filler contents, Fig. 5.22.

5.2.3.6   Temperature Increase During Curing of UP Resin

Energy is released when reactive resins cure (see also Section 1.2.3.9). To obtain some indication of the temperature rise during such a chemical reaction, the temperature of UP resins curing under pressure was measured. Relevance to practical processing parameters was ensured by pressurizing at 150 bar and an isothermal temperature of 105 °C. Figure

5.23 shows the temperature increase in the pure UP resin specimen, compared to chalk-filled batches, for curing at 105 °C. The temperature in the unfilled specimen rises by approx. 90 °C during the reaction. In the chalk-filled batches, the curing reaction was observed to shift toward longer times. The temperature rise flattens rapidly from unfilled to filled resin and with increase in filler content.

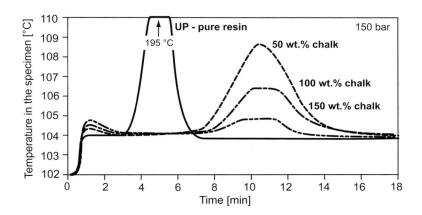

Fig. 5.23   Heating of UP resin in the pvT device during curing, as a function of chalk content

Isothermal 105 °C, pressure 158 bar, direct method

5.3   References

[1]   Rieger, J.,   Temperatur- und Druckabhängigkeit der Dichte
      Kressler, J.,   von Polymeren
      Maier, R.-D.   Personal Information, Octobre 2002

[2]   Bayer AG   Process Parameters as Production Cost Factors in the
                                     Injection Molding of Thermoplastics – Melt, Mold,
                                     Demolding Temperatures, Cycle Time, pvT-Diagram
                                     Applied Technical Information
                                     GB Kunststoffe, ATI 916

[3]   Rodgers, P. A.   Pressure-Volume-Temperature Relationships for
                                     Polymeric Liquids: A Review of Equations of State and
                                     Their Characteristic Parameters for 56 Polymers
                                     J. Appl. Polymer Sci. 48 (1993), p. 1061

[4]   Menges G.,   Das physikalische Verhalten von Thermoplasten
        Thienel, P.,   bei der Aufnahme von pvT-Diagrammen unter
        Kemper, W.   verschiedenen Meßbedingungen
                                      Plastverarbeiter 28 (1977) 12, p. 632

[5]   Schenkel, G.   Thermodynamische Grundlagen der
                                      Kunststoffverarbeitung
                                      Kunststoff und Gummi (1969) 7, p. 237 and p. 282

[6]   Hellwege, K.H.,   Die isotherme Kompressibilität einiger amorpher
        Knappe, W.,   und teilkristalliner Hochpolymere im Temperaturbereich
        Lehmann, P.   von 20 bis 250 °C und bei Drücken bis 2000 kp/cm²
                                      Z. f. Polymere 183 (1962), 110

[7]   Fuchs, K.   Entwicklung und Charakterisierung thermotroper
                                      Polymerblends
                                      Dissertation, Institut für Makromolekulare Chemie,
                                      Universität Freiburg i.Br., 2001

[8]   Menges, G.,   Eine Meßvorrichtung zur Aufnahme von p-v-T
        Thienel, P.   Diagrammen bei praktischen Abkühlgeschwindigkeiten
                                      Kunststoffe 65 (1975) 10, pp. 696–699

[9]   Stitz, S.   Analyse der Formteilbildung beim Spritzgießen von
                                      Plastomeren als Grundlage für die Prozeßsteuerung
                                      Dissertation, Institut für Kunststoffverarbeitung,
                                      RWTH Aachen 1973

[10]   Thienel, P.,   Praktische Anwendungsbeispiele für die Benutzung
        Kemper, W.,   von p-v-T Diagrammen
        Schmidt, L.   Plastverarbeiter 30 (1979) 1, pp. 22–26

Other References:

[11]   Ehrenstein, G.W.   Polymeric Materials
                                      Carl Hanser Publishers, Munich 2001

[12]   Wacker, M.,   Rheological and Thermoanalytical Investigations
        Ehrenstein, G.W.   on a "Class A" LP-SMC-Paste
                                      ANTEC 2002, San Francisco

[13]   Stitz, S.,   Spritzgießtechnik
        Keller, W.   Carl Hanser Publishers, Munich 2001

[14]   Zoller, P.,   Standard Pressure-Volume-Temperature Data
        Walsh, D.   for Polymers
                                           Technomic Publication Company Lancaster, 1995

[15]   Krause, R. D.   Modelierung und Simulation rheologisch/
                                           thermo-dynamischer Vorgänge bei der Herstellung
                                           großflächiger thermoplastischer Formteile mittels
                                           Kompressionsformverfahren
                                           VDI-Verlag GmbH, Düsseldorf 2000

[16]   Karlou, K.,   DSC and pvT Study of PVC/PMMA Blends
        Schneider, H. A.   Journal of Thermal Analysis and Calorimetry 59 (2000),
                                           pp. 59–69

6   Dynamic Mechanical Analysis (DMA)

6.1   Principles of DMA

6.1.1 Introduction

Dynamic mechanical analysis (DMA) yields information about the mechanical properties of a specimen placed in minor, usually sinusoidal, oscillation as a function of time and temperature by subjecting it to a small, usually sinusidal, oscillating force.

The applied mechanical load, that is, stress, elicits a corresponding strain (deformation) whose amplitude and phase shift can be determined, Fig. 6.1. ISO 6721-1 [1] states that the mode of deformation governs whether the complex modulus is E*, G*, K*, or L*. The other relationships are shown below for the elastic modulus E, see also ASTM D 4092 [2].

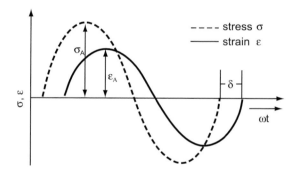

Fig. 6.1   Sinusoidal oscillation and response of a linear-viscoelastic material; δ = phase angle, E = tensile modulus, G = shear modulus, K = bulk compression modulus, L = uniaxial-strain modulus

The **complex modulus** E* is the ratio of the stress amplitude to the strain amplitude and represents the stiffness of the material. The magnitude of the complex modulus is:

$$|E^*| = \frac{\sigma_A}{\varepsilon_A}$$

The complex modulus is composed of the storage modulus E′ (real part) and the loss modulus E″ (imaginary part), Fig. 6.2. These are dynamic elastic characteristics and are material-specific; their magnitude depends critically on the frequency as well as the measuring conditions and history of the specimen.

In the linear-viscoelastic range, the stress response has the same frequency ($\omega = 2\pi f$) as the deformation input excitation. The analytical parameters in dynamic tests are the amplitudes of the deformation and the stress, and the time displacement δ/ω between deformation and stress, and are used to determine the specimen's characteristics [3].

$$|E*| = \frac{\sigma_A}{\varepsilon_A}$$

$$|E*| = \sqrt{[E'(\omega)]^2 + [E''(\omega)]^2}$$

$$E'(\omega) = |E*| \cdot \cos \delta$$

$$E''(\omega) = |E*| \cdot \sin \delta$$

$$\tan \delta = \frac{E''(\omega)}{E'(\omega)}$$

Fig. 6.2   Formulae for calculating complex modulus E*, storage modulus E′, loss modulus E″ and loss factor tan δ [1, 2]

According to ISO 6721-1 [1], the **storage modulus E′** represents the stiffness of a visco-elastic material and is proportional to the energy stored during a loading cycle. It is roughly equal to the elastic modulus for a single, rapid stress at low load and reversible deformation, and is thus largely equivalent to the tabulated figures quoted in DIN 53457.

In the same ISO standard [1], the **loss modulus E″** is defined as being proportional to the energy dissipated during one loading cycle. It represents, for example, energy lost as heat, and is a measure of vibrational energy that has been converted during vibration and that cannot be recovered. According to [1], modulus values are expressed in MPa, but N/mm^2 are sometimes used.

The real part of the modulus may be used for assessing the elastic properties, and the imaginary part for the viscous properties [3].

The **phase angle δ** is the phase difference between the dynamic stress and the dynamic strain in a viscoelastic material subjected to a sinusoidal oscillation. The phase angle is expressed in radians (rad) [1].

The **loss factor tan δ** is the ratio of loss modulus to storage modulus [1]. It is a measure of the energy lost, expressed in terms of the recoverable energy, and represents mechanical damping or internal friction in a viscoelastic system. The loss factor tan δ is expressed as a dimensionless number. A high tan δ value is indicative of a material that has a high, nonelastic strain component, while a low value indicates one that is more elastic.

In a purely elastic material (Fig. 6.3), the stress and deformation are in phase (δ = 0), that is, the complex modulus E* is the ratio of the stress amplitude to the deformation amplitude and is equivalent to the storage modulus E´ (δ = 0, therefore cosine 0 = 1; sine 0 = 0, therefore E* = E´). Steel is an example of an almost purely elastic material. In a purely viscous material, such as a liquid, the phase angle is 90°. In this case, E* is equal to the loss modulus E´´, the viscous part.

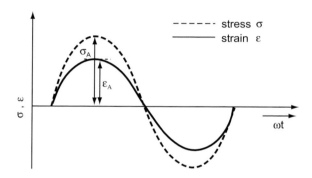

Fig. 6.3   Stress–strain behavior of a purely elastic material

Typical curves of the changes undergone by amorphous thermoplastics are shown in Fig. 6.4. At low temperatures, the molecules are so immobile that they are unable to resonate with the oscillatory loads and therefore remain stiff. The macromolecular segments cannot change shape, especially through rotation about C–C bonds, and so the molecular entanglements act as rigid crosslinks. At elevated temperatures, the molecular segments become readily mobile and have no difficulty resonating with the load. The entanglements more or less remain firmly in place, but may occasionally slip and become disentangled. Thermosets and elastomers have additional chemical crosslinks that are retained no matter what the temperature is. Weakly crosslinked rubber has one crosslink for every 1000 atoms while cured, brittle thermosets have one for every 20 atoms.

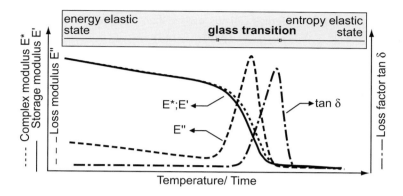

Fig. 6.4   Schematic diagram of typical DMA curves for an amorphous polymer

The material is said to be in the **glass state or energy elastic state** at the low temperatures described above, and in the **rubber or entropy elastic state** at the elevated temperatures mentioned there. A change from the glass state into the rubber-elastic state is called the **glass transition**. When the timescale of molecular motion coincides with that of mechanical deformation, each oscillation is converted into the maximum-possible internal friction and nonelastic deformation. The loss modulus, which is a measure of this dissipated energy, also reaches a maximum. In the glass transition region, the storage modulus falls during heating to a level of one-thousandth to one ten-thousandth of its original value. Because the loss factor is the ratio of the loss modulus to the storage modulus, the drop in storage modulus suppresses the rise in the loss factor initially; the temperature at which the loss factor is a maximum is therefore higher than the temperature corresponding to maximum loss modulus.

Temperature of tan δ_{max} is always higher than that of E''_{max}.

In DMA measurements, the design of the apparatus dictates that the applied loads be small. Consequently, the materials exhibit an almost purely elastic or, at least, a linear-viscoelastic response. Because the main difference between the complex modulus and the storage modulus is the nonelastic part, the smaller the nonelastic part, the smaller the difference. E^* then becomes equal to E'. Only in the glass transition, where nonelastic deformation per oscillation is a maximum, does the difference manifest itself, showing up as a decline several degrees Celsius earlier than expected.

When the results of a DMA measurement are being translated to real parts, it must always be remembered that, as the magnitude and duration of loading increase, events such as the glass transition occur a few degrees Celsius earlier than the DMA measurement indicates [4].

6.1.2 Measuring Principle

There are basically two types of DMA measurement. Deformation-controlled tests apply a sinusoidal deformation to the specimen and measure the stress. Force-controlled tests apply a dynamic sinusoidal stress and measure the deformation. Dynamic load may essentially be achieved in free vibration or in forced vibration. There are two designs of apparatus:

–   Torsion type,

–   Bending, tension, compression, shear type.

6.1.2.1   Free Vibration

In **free torsional vibration**, one end of the specimen is clamped firmly while a torsion vibration disc at the other end is made to oscillate freely. The resultant frequency and amplitude of the oscillations, along with the specimen's dimensions, are used to calculate the torsion modulus. Measurements are conducted at various temperatures to establish how the torsion modulus varies with temperature. The term torsion modulus is intended to convey the idea that the stress is not necessarily purely shear and that the observation is not necessarily a shear modulus (except in the case of cylinders). On being twisted, the flat, clamped specimen is placed in torsional stress and, to an extent depending on the way it is clamped and on its shape, its two free edges are placed in tension and its center is placed in compression.

Another free-vibration method is flexural vibration. In this, the specimen is firmly clamped between two parallel oscillation arms. One arm keeps the specimen oscillating so that the system attains a resonance frequency of almost constant amplitude. The modulus is calculated from the resonance frequency, the resultant amplitude, and the dimensions of the specimen.

Free-vibration apparatus (resonant) is highly sensitive and is eminently suitable for studying weak effects. The disadvantage is a drop in frequency combined with falling modulus due to elevated temperatures [5]. Frequency-dependent measurements are difficult to perform and require the use of different test-piece geometries. They thus entail considerable experimental outlay.

For more information see also ASTM D 5279 [6], ISO 6721-7 [7], ISO 6721-2 [8] and ISO 6721-3 [9].

6.1.2.2   Forced Vibration (Non-resonant)

Variable frequency apparatus applies a constant amplitude (stress or deformation amplitude). The frequency may be varied during the measurement. Figure 6.5 shows the design of a torsion apparatus.

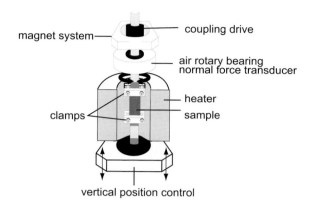

Fig. 6.5   Schematic design of a torsion vibration apparatus with variable frequency [10]

Firmly clamped at both ends, the specimen is electromagnetically excited into sinusoidal oscillation of defined amplitude and frequency. Because of the damping properties of the material or specimen, the torque lags behind the deformation by a value equal to the phase angle δ, see Fig. 6.1. The observed values for torque, phase angle, and geometry constant of the specimen may be substituted into the formulae listed above to calculate the complex modulus G*, the storage modulus G′, the loss modulus G′′, and the loss factor tan δ.

The specimen should be dimensionally stable and of rectangular or cylindrical cross section. Suitable specimens have modulus values ranging from very high (fiber composites) down to low (elastomers). If appropriate plane-parallel plates are attached to the drive shafts, it is also possible to measure soft, gelatinous substances and viscous liquids [11, 12].

Most types of apparatus utilize vertical loading, which allows measurements under bending, tension, compression, and shear. Usually, the same apparatus is employed, with inter-changeable clamping mechanisms applying the various types of load, Fig. 6.5.

In **three-point bending**, the ends are freely supported and the load is applied to the midpoint, [13, 14]. To ensure direct contact with the specimen, an additional inertial member needs to be applied. This test arrangement is suitable for very stiff materials, such as metals, ceramics, and composites. It is unsuitable for amorphous polymers because they soften extensively above the T_g. It is a simple arrangement, but additional shear stress in the midpoint plane of the specimen must be taken into account. With short specimens, this gives rise to interlaminar shear stress in the neutral, usually shear-soft plane. The effect can be reduced by employing either an appropriate length/thickness ratio or a four-point arrange-ment, which is usually more complicated.

Provided that bending specimens are firmly clamped, it is also possible to measure amorphous thermoplastics above the T_g. The specimen is clamped to both supports and, at its midpoint, to the push-rod (dual **cantilever bending** stress) [15]. Consequently, no inertial member is needed. This test arrangement is used for reinforced thermosets, thermoplastics, and elastomers. Specimens that expand considerably when heated may distort in a dual cantilever arrangement, and this can falsify the readings. For such specimens, a single-cantilever or freely supported arrangement is best. While a specimen experiences alternate compression and tension mainly along its length in flexural loading, it experiences homogeneous stress down its longitudinal axis when either tension or compression is applied exclusively.

Tensile mode is ideal for examining thin specimens, such as films and fibers in the low-to-medium modulus range. Clamped at top and bottom, the specimen is subjected to an underlying tensile stress to prevent it from buckling during dynamic loading [16, 17].

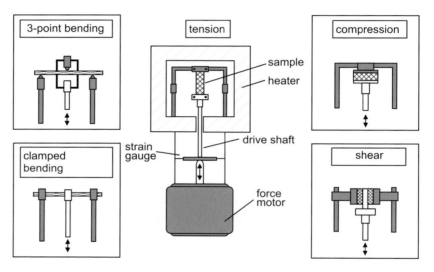

Fig. 6.6   Schematic design of a dynamic-mechanical analyzer under vertical load, and showing the various possible test arrangements

In **compression mode** an axial load is applied to mostly specimens held between two parallel plates [18]. Soft rubbers on gelations pastes are most suitable for this measurement. Uniaxial compression effects a one-dimensional change in the specimen's geometry whereas bulk compression effects a three-dimensional change. Uniaxial compression can cause thin specimens to buckle. On short, thick specimens, hindered deformation at the supports makes it difficult to make accurate determination of the modulus.

Like compression loading, axial **shear loading** is suitable for soft materials [19]. Good results are produced by a sandwich arrangement in which two specimens are subjected to cyclical shear by the displacement of a central push-rod.

6.1.3 Procedure and Influential Factors

The stages involved in a DMA measurement are as follows:
- choose a load appropriate to the problem, and a clamping device,
- prepare specimen (geometry, degree of plane-parallelism),
- clamp the specimen,
- choose measuring parameters.

Factors exerting an influence on the apparatus and specimen are:

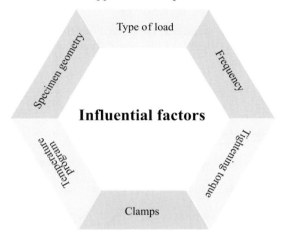

The influential factors and sources of error associated with experiments are explained in detail in Section 6.2.2 with the aid of curves plots from real-life examples.

6.1.4 Evaluation

Because DMA is sensitive to variations in the stiffness of a material, it may be used to determine not only modulus and damping values directly, but also glass transition temperatures. It is particularly suitable for determining glass transitions because the change in modulus is much more pronounced in DMA than, for example, the c_p change in a DSC measurement [20]. Owing to discrepancies between the proposals made in various standards and the information provided by apparatus manufacturers, confusion has arisen about how to determine and state glass transition temperatures in practice. Although the modulus step that occurs during the glass transition can be evaluated much in the manner of a DSC curve, it is difficult to do so in practice. This and other methods of determination are described below.

Methods of determining the glass transition temperature:

Evaluation of modulus step:
- Step method employed for DSC curves,
  (start, half step height and end of glass transition),

- Inflection point method,
- 2% offset method as set out in [21]
  (start of glass transition),
- Tangent method as set out in [21] (start of glass transition).

Evaluation of peaks from plots of loss factor and loss modulus:
- Maximum loss factor,
- Maximum loss modulus.

6.1.4.1   Methods of Evaluating the Modulus Step

The use of the modulus step to determine the glass transition temperature is based on the standardized DSC method (ISO 11357-1 [22], Fig. 6.7) and involves ascertaining the onset, end, and midpoint temperatures.

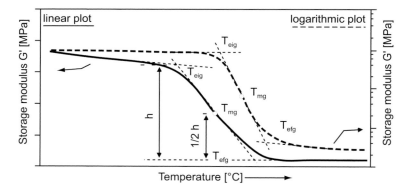

T_{eig}	Extrapolated onset temperature, onset temperature:	Intersection of the inflectional tangent with the tangent extrapolated from temperatures below the glass transition
T_{mg}	Midpoint temperature, glass transition temperature:	Temperature of the midpoint of the inflectional tangent (half step height between T_{eig} and T_{efg}), projected onto the DMA curve
T_{efg}	Extrapolated end temperature:	Intersection of the inflectional tangent with the tangent extrapolated from temperatures above the glass transition

Fig. 6.7   Step evaluation based on the standardized DSC evaluation

T_{eig} = Extrapolated onset temperature
T_{mg} = Midpoint temperature
T_{efg} = Extrapolated end temperature, linear and logarithmic plots of modulus

Tangents are applied to the sections of the curve above and below the glass transition step. An inflectional tangent applied to the step intersects with both these tangents at the extrapolated onset temperature T_{eig} and the extrapolated end temperature T_{efg}. The midpoint temperature T_{mg} is determined from the half step height.

The user chooses the temperatures at which to apply the tangents to be determined – this practice differs from that described in [21]. A regression curve is then calculated from values above and below these temperatures. The position of the resultant characteristic curve depends critically on how the modulus is plotted against the temperature. For step evaluation methods, logarithmic plots yield much higher T_g values than linear plots.

These two different ways of plotting storage modulus have their advantages and disadvantages:

	Logarithmic plot	Linear plot
Below T_g:	– Seemingly slight, often linear dependence on temperature – Tangents usually easy to apply	– Clear, heavy dependence on temperature – Difficult to apply tangents, especially with nonlinear curves – Choice of contact temperatures for the tangents on the curve is subjective
	Logarithmic plot	Linear plot
Above T_g:	– Mathematical stretching and thus steeper curves – Tangents difficult to apply	– As for logarithmic plot below T_g
T_{mg}:	– Half step height is a linear measurement in a logarithmic plot: not appropriate	– Half step height appropriate; founded more on physical structure than on technical design reasons.
	T_{mg}	>	T_{mg}

Logarithmic plot of modulus: T_{mg} higher – subjective
Linear plot of modulus: T_{mg} lower – appropriate

Note: *Most users are interested in knowing the highest service temperature of a particular polymer. Generally the extrapolated onset temperature may be quoted for this. As the interpretation of this is partly subjective, it is preferable to use the glass transition temperature defined, in, for example, standards and procedures. Note, however, that the polymer often cannot be used at this temperature in practice.*

In the turning point method, the glass transition is identified by mathematically identifying the **turning point (point of inflection)** of the modulus step, Fig. 6.8. This is most often done by calculating the first derivative of the curve. But even this method can produce different T_g values because software programs use different algorithms for the calculation.

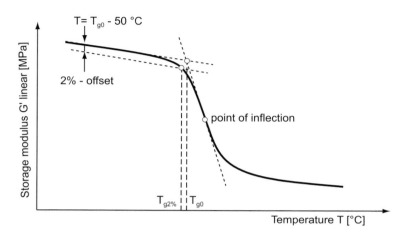

$T_{g2\%}$	*Start of the glass transition by the 2% method*
T_{g0}	*Start of the glass transition by the tangent method*
2-% drop	*2-% drop in modulus starting from the modulus at $T = T_{g0} - 50\ °C$*

Fig. 6.8   Determination of the glass transition (2% method) from the storage modulus-temperature curve as set out in DIN 65 583 (April 1999) [21]

Evaluation in accordance with DIN 65 583 (Draft, 1990); see Fig. 6.27 and Fig. 6.28

DIN 65 583 [21] describes two methods for determining the temperature at the start of the glass transition that differ from the stepwise evaluation already described. In the **tangent method**, tangents are applied to the linear portion of the curve of storage modulus against temperature below the glass transition and to the inflection point of the rapid drop in storage modulus. The temperature at the intersection of the tangents is defined as T_{g0}.

As an alternative to the tangent method, [21] use the **2% method** for fiber-reinforced polymers. In [21], a line is drawn parallel to the tangent on the linear portion of the curve of storage modulus against temperature at a temperature of (T_{g0} -50 °C), expressed in terms of the storage modulus, that is 2% below the tangent. The intersection of these parallels with the curve of the storage modulus is defined as the start of the glass transition $T_{g2\%}$, Fig. 6.8. This value is suitable in parts design to describe the limit of thermal use, that is, the start of softening.

Where the storage modulus curve is curvilinear, it is difficult to apply a defined tangent in the energy-elastic zone ($< T_g$). Evaluation is made easier by specifying a fixed temperature for applying the tangent, as set out in the draft DIN 65 583 standard from 1990, Fig. 6.27, Fig. 6.28.

The temperatures at which the tangents are applied are sometimes specified by standards or are chosen by the user.

6.1.4.2   Evaluation of Peaks

The glass transition temperature is often taken to be the temperature of the **maximum loss modulus** (E''_{max} or G''_{max}) or the **maximum loss factor** (tan δ_{max}). Such curves are easier to evaluate than step curves. Fig. 6.9 shows how the various methods compare.

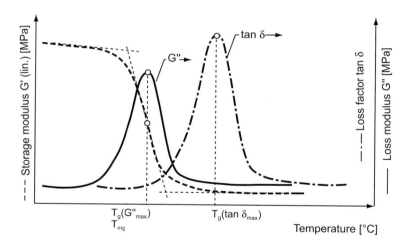

Fig. 6.9   Glass transition temperature as the maximum loss modulus G''_{max} and maximum loss factor tan δ_{max} as opposed to the midpoint temperature T_{mg} used in step evaluation

The maximum loss factor yields the highest glass transition temperatures, higher than any T_g value discussed so far.

Rieger [23] believes there are several arguments in favor of defining the glass transition temperature in terms of the maximum loss modulus G''_{max} instead of the maximum loss factor tan δ_{max}. Because G'' is a measure of dissipated energy, he contends that it makes sense to define the temperature at maximum loss (G''_{max}) as the transition temperature. A further argument is that the temperature of maximum G'' remains the same, regardless of whether a pure material or a blend is being considered [23].

ASTM D 4065-2001 [24] also recommends that the evaluation be based on the temperature of maximum loss modulus.

Determining the glass transition temperature from the maximum loss modulus is fairly straightforward. Furthermore, the value agrees well with the temperature given by DMA step evaluation (linear plot, half height). Problems can arise, however, if the loss modulus maximum is not sufficiently accentuated.

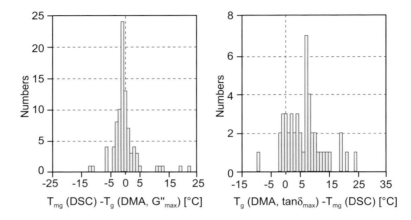

Fig. 6.10   Frequency distribution of the temperature differences for various methods of
              determining T_g [23]

T_{mg} (DSC) = midpoint temperature
T_g (DMA, $\mathbf{G''_{max}}$) = Loss modulus maximum
T_g (DMA, tan δ_{max}) = Loss factor maximum

Generally, the temperature of G''_{max} at a frequency of 1 Hz corresponds to the midpoint temperature T_{mg} (half step height) observed in DSC [23]. A large number of experiments on 40 representatives of different classes of polymer have borne out this observation. These compared the frequency of the temperature difference between T_{mg} (DSC) and T_g (DMA, G''_{max}) and between T_{mg} (DSC) and T_g (DMA, δ_{max}), Fig. 6.10. The DSC measurements were performed at a heating rate of 20 °C/min and as set out in [25], while the DMA studies involved a torsion pendulum and a heating rate of 1 °C/min.

These distributions mostly reveal very good agreement between T_{mg} (DSC) and T_g (DMA, G''_{max}). In isolated cases, however, there were differences of ± 10 °C and more. There is much less correlation between T_{mg} (DSC) and T_g (DMA, tan δ_{max}) [9]. We were unable to verify as more correct the assertion that the DMA glass transition temperature, defined as the mean value of the peaks of E'' and tan δ, differs by no more than ± 4 °C from the glass transition temperature T_{mg} observed in DSC [26].

Peaks are relatively easy to evaluate.
$$\mathbf{T_g \, (tan \, \delta_{max}) \, always > T_g \, (G''_{max})}$$
$$\mathbf{T_g \, (G''_{max}) \approx T_{mg} \, (step, \, linear) \approx T_{mg} \, (DSC)}$$

Elastomers are characterized by means of the low temperature reference point T_R , determined as the maximum on the loss modulus curve [27].

In summary, it may be said that different methods of determining T_g yield different values for T_g. When a glass transition temperature is stated, therefore, it is absolutely vital to indicate the method of evaluation in addition to the experimental parameters (see Section 6.2.2). See also ASTM E 1640 [28].

When stating T_g values, describe the evaluation method and measuring parameters.

The various evaluation methods will be illustrated again on real-life examples in Section 6.2.2.7.

6.1.4.3   Test Report

ISO 6721-1 [1] contains valuable information for compiling a complete test report describing all experimental parameters and specimen information. The test report should include the following information, where appropriate:

- A reference to standards employed.
- All details necessary for complete identification of the material tested.
- For sheets, the thickness of the sheet and, if applicable, the direction of the major axes of the specimens in relation to some feature of the sheet, for example, direction in which sheet is processed.
- Shape and dimensions of the specimens.
- Method of preparing the specimens.
- Details of the conditioning of the specimens.
- Number of specimens tested; details of the test atmosphere if air is not used.
- Description of the apparatus used for the test, and the test conditions (frequency, type of load, load amplitude etc.).

- Temperature program used for the test, including the starting and final temperatures as well as the rate of linear change in temperature or the size and duration of the temperature steps.
- Table of data.
- DMA curves.
- Date of test.

6.1.5 Calibration

There is no standardized procedure for calibrating DMA apparatus. Several standards merely make recommendations. Instruments should be calibrated in line with the manufacturer's procedures. Temperature and modulus also need to be calibrated. Regular calibration ensures that observations will remain comparable over time.

6.1.5.1   Temperature Calibration

During a measurement, discrepancies arise between the temperature displayed on the apparatus and the specimen's real temperature. These are due first to differences in the thermal conductivity and heat capacity of different specimens and second to dissipation of heat via the clamps for holding the specimens on the apparatus. Sections of specimens close to the clamps usually do not attain the specified specimen temperature, that is, the temperature displayed is higher than that in the specimen. The consequence of this is that the specimen is observed to be stiffer at this temperature than it actually is [29].

It is rarely possible to measure the real temperature of the specimen because the apparatus would be too complex. Using the melting points of metals to calibrate the temperature is not wholly satisfactory either. Many systems attempt to compensate for the difference by completely enveloping the temperature sensor, but this causes a timelag in the temperature display.

It therefore makes sense to regularly compare observed transition temperatures with observations from reference specimens. DIN 65583 [21] recommends polycarbonate specimens for temperature corrections – the observed damping maximum is compared with the literature value of 153.5 °C [21] and the observations are corrected accordingly.

ASTM D 4065-2001 [24] recommends water or indium for calibrating the temperature. Other common calibrating agents are metals encapsulated in an insulating layer of polymer. ASTM E 1867 [30] describes how to perform a temperature calibration using materials of known melting point. These materials are either liquids or metals that are also used for calibrating other thermoanalytical instruments. In accordance with their physical state, they are either introduced into a PEEK capsule or wrapped in aluminium foil. This "package" is then placed in the DMA and measured under the appropriate conditions. Melting of the calibration medium gives rise to a step in storage modulus in the DMA. The start of this step is evaluated by the tangent method and used to create a linear calibration plot of storage modulus.

The basic requirement is that the calibration conditions must be exactly those used for the experiment. See also ASTM E 1867 [30].

> **Calibration conditions = Experimental conditions**

6.1.5.2   Modulus Calibration

The modulus is usually calibrated with the aid of defined steel or aluminum specimens. For polymers, it is best to calibrate the modulus with the aid of standard specimens whose stiffness you have determined accurately by means of other methods. See also ASTM E 2254 [31].

6.1.5.3   Apparatus Calibration

Many types of apparatus need to be calibrated according to the manufacturer's procedures (stiffness, yieldingness, damping, and moment of inertia of the oscillation system). Keeping records of daily calibration values allows gradual changes to be detected that may be caused by contamination of the air bearing, wear, or maladjustment of the drive shaft, and so forth.

6.1.6   Overview of Practical Applications

Table 6.1 shows which DMA characteristics can be used to describe quality defects, processing flaws, and other parameters. These are treated in more detail in Section 6.2.3 Real-Life Examples.

Application	Characteristic	Example
Regions in which state is dependent on temperature	E'	Energy and entropy-elastic region, start of melting
Temperature-dependent stiffness	E', E'', T_g, tan δ	Elastic and nonelastic response
Thermal limits on use	T_g	Start of softening or embrittlement
Frequency- and temperature-dependent damping	tan δ (f)	Response of damping elements
Blend of constituents difficult to identify by DSC	T_g	Impact-modification of Polyamid 6 through butadiene rubber

Application	Characteristic	Example
Influence of fiber reinforcement on mechanical parameters	E', E'', tan δ	Anisotropic stiffness
Recycling, repeated processing, aging	T_{g1}, T_{g2}	Shift in butadiene T_g from ABS to higher temperatures
State of aging (conditioning)	T_g	Water content of PA
Degree of curing, postcuring	T_g	T_g rises, tan δ falls, modulus rises
Thermal degradation	T_g	T_g falls

Table 6.1　　Some practical applications of DMA experiments on plastics, along with relevant characteristics

6.2   Procedure

6.2.1   In a Nutshell

Mode of deformation

Torsion, bending, tension, compression or shear, the choice depending on the purpose of the measurement, the specimen's geometry, and its consistency.

Load amplitude/ deformation

In all measurement methods, the loads are far smaller than those experienced by real-life parts. Consequently, the observed viscoelastic and viscous deformation components are much lower. For the measurement, the deformation may be specified as deflection (μm, °, %) or force or torque. The specified deformation must lie in the viscoelastic range.

> **Deformation during measurement must remain in the linear viscoelastic range.**

Test arrangement/ specimen geometry

Torsion: Long specimens of rectangular or circular (cylindrical) crosssection and high modulus; Plate/plate arrangement: Soft, paste-like specimens.

Three- or four-point bending: Specimens of high modulus (fiber-reinforced)

Dual cantilever bending: Specimens of low-to-medium modulus

Single cantilever bending: Specimens of medium modulus and high thermal expansion

Tension: Films, microtome sections, fibers, rodlike specimens of low crosssectional area

Compression: Specimens of low modulus

Shear: Soft specimens

Specimen preparation

Injection-molded, milled, punched, or sawn plane-parallel or circular specimens with no flash, notches, or sink marks. When sampling, take into account direction of reinforcing fibers, direction of processing, and state of aging. Do not expose specimens to thermal influences.

Clamping/grips   The clamping mechanism is dictated by the test arrange-
                            ment, and the type of grip used depends on the specimen.
                            It is vital to ensure the grips are evenly mounted so that
                            the specimen is not constricted or deformed. Friction must
                            not occur between the specimen and grip.

Purge gas   An inert purge gas may be used for preventing oxidation.

Measuring program   The measuring program is used to specify starting and
                            finishing temperatures, heating rate, frequency, and
                            deformation.

                            The starting temperature should be 30–50 °C below the
                            temperature of the expected event. It must be kept constant
                            until temperature equilibrium has been reached throughout
                            the entire specimen.

                            As the specimen crosssection is relatively large, the
                            maximum heating rate is 1–5 °C/min (heating $\approx$ wall
                            thickness^2).

> **The best heatingrate is < 3 °C/min.**

                            The maximum finishing temperature is governed by
                            melting in the case of thermoplastics or incipient decom-
                            position of thermosets.

Frequency   For nonresonant oscillation, the frequencies are varied
                            mostly incrementally between 0.001 and 200 Hz, the
                            maximum frequency being lower than the resonance
                            frequency of the specimen/apparatus. Measurements may
                            be performed at a single frequency (standard frequency is
                            ~ 1 Hz) or multiple frequencies.

> **Standard frequency ~ 1 Hz**

Evaluation   Change of modulus (complex modulus, storage modulus,
                            loss modulus), mechanical loss factor, and T_g. Evaluation
                            affected by linear and logarithmic plots.

Interpretation   Information about temperature-dependent change of stiffness, glass transition region, viscoelastic response. Specimen parameters, measuring parameters and evaluation routines must be borne in mind.

Note: **Elastic** = *Load and deformation are directly proportional. No irreversible damage or changes in material occur; characteristic feature: linear stress–strain diagram (v = const.)*
Linear-viscoelastic = *Delay between load and deformation. Inelastic deformation components are proportional to stress, temperature and time; characteristic feature: linear isochronous stress–strain diagram (t = const.)*
Nonlinear-viscoelastic = *Inelastic deformation components depend disproportionately on the stress and temperature;*
Characteristic feature: No linearity between stress and strain.

6.2.2   Influential Factors and Possible Errors During Measurement

6.2.2.1   Mode of Deformation

Conventional apparatus permits measurement in torsion, bending , tension, compression and shear. The loads which can be applied are relatively low, with torques of around 1.5 Nm or forces of about 18 N (apparatus capable of higher forces, for example, hydropulsers, is not discussed here). In real life, the loads are frequently higher and occur in combinations.

The least amount of force to generate deformation is needed for stiff or reinforced materials tested in bending mode. If the specimen is restricted in terms of thickness or length, the issue of clamping becomes important. The closer the stiffness of the specimen resembles that of the measuring system, the poorer is the signal/noise ratio and the less meaningful are the results.

Specimen geometry influences the quality of measurement.

A handy way of establishing the best specimen geometry for the deformation involved is to use a **geometry factor k** to represent the influence of the geometry in different test arrangements. It is given by the modulus (G, E), the force (F) or the torque (M), the deflection (φ, Δx, Δl) and the geometry of the sample, diameter (d), length (l), width (b), height (h) and cross-secional area (A) and may be determined as follows:

Torsion: $\qquad G \cdot k = \dfrac{M}{\varphi} \qquad k_o \approx \dfrac{d^4}{l}$

Bending : $\qquad E \cdot k = \dfrac{F}{\Delta x} \qquad k \approx \dfrac{b \cdot h^3}{l^3}$

Tension/compression: $\qquad E \cdot k = \dfrac{F}{\Delta l} \qquad k \approx \dfrac{A}{l}$

Shear: $\qquad G \cdot k = \dfrac{F}{\Delta x} \qquad k \approx \dfrac{A}{b}$

The geometry factor and the expected modulus can be used to estimate whether the apparatus can generate the range of forces necessary for the desired mode of deformation, and whether the specified force produces a measurable deflection of the specimen. The brief overview of polymers and their properties in Chapter 8 lists several elastic modulus values [32] that can serve as a guide.

Note: *Modulus values may vary considerably according to manufacturer and processing conditions. G moduli are often measured in torsion and thus do not constitute unequivocal shear moduli. For this reason, they are often also called torsion modulus, but this is not a standardized term.*

The moduli of elasticity (E) and shear (G) observed in conditions of unequivocal stress states are related by Poisson's ratio μ : $E = 2G(1+\mu)$. For polymers, μ is usually between 0.33 and 0.45. The upper value applies to elevated temperatures, high stress and longer duration.

6.2.2.2   Load Amplitude/Deformation

The quality and the accuracy of a DMA measurement are critically dependent on the level of deformation needed for loading the specimen. Apparatus is designed to measure either the force or torque needed to produce a specified deflection (μm, °, %) or the deflection produced by a specified force or torque.

Irrespective of apparatus design, the deformation must lie within the linear viscoelastic range of the material because otherwise damage-free measurement cannot be ensured and the stress–strain response will not be described accurately by the mathematics. For solid polymers, the deformation must not exceed several tenths of 1% [33]. Although apparatus designed for measuring these materials generally does not allow extensive deformation, check that the chosen deformation lies in the linear viscoelastic range or else find a more suitable mode of deformation.

The linear viscoelastic response of a system in deformation has to be established experimentally. This can be done by measuring the moduli as a function of deformation at a fixed frequency [34]. Provided the storage and loss moduli both remain linear with change in stress and deformation at a certain frequency and temperature, the material is only experiencing linear, viscoelastic stress, Fig. 6.11. If several frequencies are being measured, perform this test at the minimum and maximum frequencies that will occur.

Deformation must lie within the linear viscoelastic range

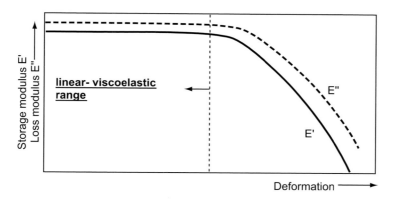

Fig. 6.11   Establishing the linear viscoelastic range by plotting storage and loss moduli as
a function of deformation

6.2.2.3   *Test Arrangement/Specimen Geometry*

This section deals with conventional test arrangements and specimen geometries, and suitable types of material for each.

Torsion

In the forced-vibration torsion pendulum test, the specimen is clamped between two grips or applied as a viscous paste between two parallel plates and subjected to a torque applied via a drive shaft. The torsion pendulum test is suitable for testing materials over a wide modulus range.

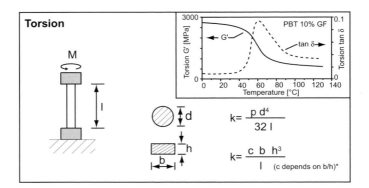

quod vide [35]

Fig. 6.12   Test arrangement, geometry factors, and typical curve for torsion

Solid specimens show a linear increase in deformation, and thus stress, from the center toward the edges. Cylindrical specimens exhibit virtually uniform stress across the cross-section but are rarely encountered in practice because they are complicated to make. Flat, rectangular specimens are the most common because they are easy to produce.

Rodlike specimens: Hard, stiff – high modulus specimens

Note: *The accuracy of the observed modulus depends on the ratio of the specimen's width to its thickness. Flat specimens are easier and faster to heat but tend to twist. Tension develops in the outer regions and pressure builds up in the center, and a spuriously high modulus is observed. Whether a particular width/thickness ratio for a specimen is favorable depends on the apparatus and the clamping geometry. Good results have been obtained with a ratio of 3:1 for thermoplastics and elastomers, and of up to 10:1 for reinforced laminates.*

The ratio of specimen width to specimen thickness is important.

Viscous fluids, such as melts and reactive resins, are measured between two plates 1–2 mm apart (parallel plate mode). To prevent soft, compact specimens from sliding over the plates, it is sometimes expedient to bond them to the plates. In that case, remember that the adhesive influences the observations.

Parallel plate mode: Melting, curing – liquid to soft-solid specimens

---

Bending

Bending tests with their different clamping methods – freely supported or gripped – offer a wide range of measurements. All sorts of grips are available to suit specimen type and geometry. In **three-point bending**, the specimen rests freely on two fixed supports which are slightly rounded to avoid stress concentrations and notching. The load is applied to the center of the specimen by a slightly rounded push-rod attached to the drive shaft. It is crucial for the specimen to be supported or clamped symmetrically.

An additional inertial member ensures close contact.

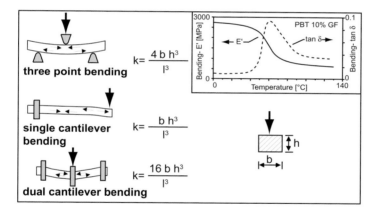

Fig. 6.13   Test arrangement, geometry factors, and typical curve for different bending modes

Rectangular crosssections are best as the specimens are uniformly supported across their width. In the three-point bending test, the ratio of the distance between the supports to the thickness of the specimen governs the shear stress in the specimen's midplane. The smaller the ratio, the higher is the shear stress.

Bending is the primary test for stiff specimens because deformation is easy and generates large signals. For specimens with a high elastic modulus, the specimens themselves should be thin and the distance between the supports should be ten times the thickness. Specimen width is only of major importance for fiber composites; a large width can offset inhomogeneities in the material.

Unfilled, amorphous thermoplastics undergo extensive deformation during softening and so it is difficult to conduct measurements beyond their glass transition region.

Three-point bending: Fiber-reinforced or highly filled polymers of high modulus, amorphous thermoplastics < T_g

Further scope for measurement is afforded by firmly clamping the specimen at one or both ends (single- or dual-cantilever bending). As the specimen is firmly clamped, no inertial member is needed. This means that amorphous thermoplastics can also be measured beyond their glass transition region.

**Cantilever bending: Unreinforced or slightly reinforced polymers
– low to medium modulus specimens
Measuring range: Above and below T_g**

Note: *Three-point bending and, to a greater extent, cantilever bending does not generate unequivocal stress states. Bending characteristics are not used for design purposes and have largely been deleted from the Campus database. They are attractive because they are easy to measure and because bending stress occurs frequently in plastic parts, which are mostly thin-walled in real life. With soft specimens, squeezing by the grips falsifies the thickness.*

Metal disks bonded to grip ends can reduce their influence, especially on soft materials.

Cantilever bending: Reduce gripping effects by using longer, thinner specimens.

If the specimen is firmly clamped at both ends, high thermal expansion along its axis may cause it to warp; this will generate additional compressive force (Fig. 6.14) and a spurious increase in stiffness that disappears again due to softening when the glass transition is passed. In cases like that, single-cantilever bending is preferable because the drive shaft can be easily pushed aside to prevent additional loading due to expansion (Fig. 6.14).

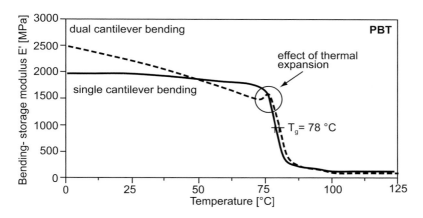

Fig. 6.14   Influence of thermal expansion when PBT is clamped in different ways

Heating rate 5 °C/min, frequency 1 Hz

High thermal expansion: Single-cantilever clamping

Tension

In tension, the specimen is held between grips on the fixed frame and the movable drive shaft, Fig. 6.15. In compression, the sample is prevented from buckling during dynamic loading by providing an inertial member.

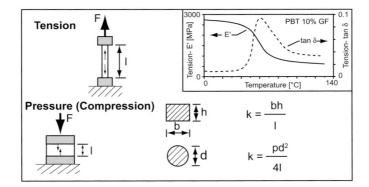

Fig. 6.15   Test arrangement, geometry factors, and typical curve for tension and compression

Decreasing the crosssection or employing a longer specimen reduces the geometry factor and accentuates the modulus signal. However, irregularities in specimen geometry become more pronounced at small specimen crosssections. Consequently, if the geometry has to be modified, it is better to increase the length than to decrease the crosssection.

**Linear influence of thickness, width and length;
rectangular crosssection and long specimen where possible;
width:length ratio of 1:3 is best**

Measurements in tension are especially recommended for films, flat specimens, fibers, and medium-modulus specimens. High-modulus specimens may be measured if the cross-section is reduced accordingly. Possible geometries for the tensile mode vary from apparatus to apparatus. Commonly employed geometries are: thickness 5 μm–2 mm, width < 10 mm, unclamped length < 30 mm.

Tension: Suitable for films, fibers, medium-modulus specimens

Note: *Tension is a simple, clear mode of deformation, in which the load is applied along the specimen axis and the stress is distributed uniformly across the crosssection. For this reason, tensile measurements usually return the most appropriate modulus values.*

Compression

In a compression measurement, the specimen is located between two parallel plates, one of which is movable (Fig. 6.15). The clamping mechanism governs the geometry of the specimens that can be measured. They are usually short and either cylindrical or rectangular, with a maximum diameter or length of 15 mm. Compression is usually performed on low-modulus materials, such as foams and elastomers.

Compression: Suitable for foams, elastomers, low-modulus specimens

Note: *In the case of short, large crosssection specimens, friction between specimen and support can lead to hindered transverse deformation. It is important for the supports and especially the specimen to be parallel as otherwise loading will not be uniform.*

Shear

In shear mode, two disc-shaped specimens are clamped symmetrically between two stationary plates and a center plate joined to a drive shaft. Oscillation of the center plate generates shear on the specimens.

Specimen geometry and modulus are contingent on each other. To measure somewhat stiffer specimens, it is best to minimize the geometry factor, that is,, to increase the specimen thickness b or reduce the crosssectional area. A specimen thickness of around 1 mm has proved useful in practice.

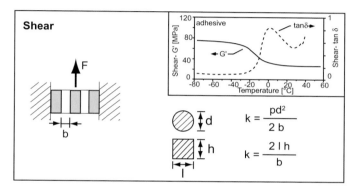

Fig. 6.16   Test arrangement, geometry factors and typical diagram for shear

Shear tests are best conducted on low-modulus materials, pastes, and viscous fluids. Brittle solids cannot be measured.

Shear: Suitable for elastomers, adhesives, viscous fluids, low-modulus specimens

Note: *Shear measurements can be used to monitor the progress of curing in adhesives and reactive resins. The end modulus value in such a system generally far exceeds the measuring scope of this test arrangement. The maximum attainable modulus is governed by the apparatus stiffness and the geometry factor. For an apparatus with a stiffness ($F/\Delta x$) of approx. 106 Pa or 1 N/mm², the maximum measuring range (see Section 6.2.2.4) for a 1-mm-thick SMC specimen of a certain diameter is given by $G_{max} = F/\Delta x \cdot k$:*

$$\textit{Specimen diameter 7 mm} \quad \Rightarrow k = 0.077 \quad \Rightarrow G_{max} \sim 13 \;\; MPa$$
$$\textit{Specimen diameter 12 mm} \quad \Rightarrow k = 0.226 \quad \Rightarrow G_{max} \sim 4.4 \; MPa$$

Adjust crosssectional area and specimen thickness to the modulus value.

Figure 6.17 illustrates this situation for a reactive SMC paste. The maximum observable value in shear for the specimen diameter agrees with the calculated values, yet still differs greatly from the actual value after the curing reaction is complete (approx. 1500 MPa). Therefore, only the start of the reaction or the gel time can be measured.

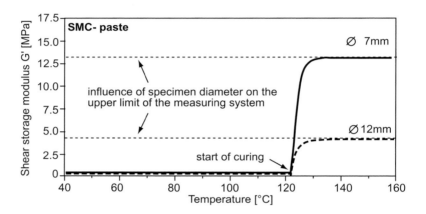

Fig. 6.17   Influence of specimen diameter on the upper limit of the measuring system; start of curing of an SMC paste

Two specimens, thickness 1 mm, shear, heating rate 2 °C/min, frequency 1 Hz

Crosslinking reaction: Start of reaction and gel time

6.2.2.4   *Specimen Preparation/Clamping*

Injection molded tensile rods or plates make good specimens. Gentle preparation, for example, milling or sawing (water-cooled), is a prerequisite for obtaining the desired specimen geometry. According to ISO 6721-1 [1], deviations in the geometry must not exceed 3% of the mean value. An 8-mm-wide specimen must therefore fluctuate in width by no more than ± 0.24 mm along its entire length. Width, thickness, and length are measured at room temperature to an accuracy of ± 0.5% with a micrometer screw gage or caliper. Thin films are punched out with an appropriate punch; soft, rubberlike specimens (e.g., elastomers) are milled frozen. Specimens with flaws may be used only for determining characteristics that are not geometry-dependent (temperatures or change in tan δ).

Avoid deviations in specimen geometry > 3%

The properties of fiber-reinforced materials are governed by the **fiber orientation**. Fig. 6.18 shows the influence of the mode of deformation on the modulus of dry CF-EP laminates as a function of temperature for different fiber orientation. Unidirectional laminates are highly anisotropic because their strength and stiffness are a maximum under tension, compression, and bending along the fiber (0° laminate) but are simultaneously a minimum at right angles to the fiber direction (90° laminate). Quasi-isotropic laminates show maximum stiffness in torsion because the 45° layers favor this mode. This contrasts with laminates whose fibers are aligned axially and circumferentially, whose stiffness is a minimum, and consequently the matrix properties dominate under torsion. A quasi-isotropic laminate with 16 layers was also studied, whose layers had the following arrangement 2 x [0°/90°/+45°/-45°/0°/ 90°/+45°/-45°]. The extrapolated onset temperatures T_{eig} shown in the diagrams differ because the fiber orientations are different. The effect of matrix softening as a function of temperature is masked to a degree dependent on the fiber orientation.

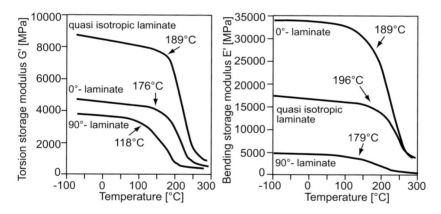

Fig. 6.18   Influence of the mode of deformation on the change in modulus and the extrapolated onset temperatures of CF-EP laminates [29]

Left: Torsion

Right: Three-Point bending, heating rate 3 °C/min, frequency 1 Hz

Modulus values and temperatures depend on the fiber orientation and mode of deformation.

The specimen must be clamped as uniformly as possible and must not undergo deformation or distortion. Special clamping devices and engraved grips are available to make uniform alignment easier.

If the starting temperature is much lower than room temperature and if the **clamped** specimen contracts more at room temperature and during subsequent cooling than the materials used for most metallic grips, the grips must be retightened at the low starting temperatures, Fig. 6.19.

Tighten grips again when starting temperatures are low.

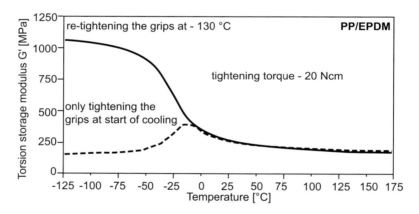

Fig. 6.19   Influence of retightening the grips at low starting temperatures on the modulus of a
               PP/EPDM specimen

               Torsion, heating rate 5 °C/min, frequency 1 Hz

The influence of the tightening torque on the modulus is shown in Fig. 6.20.

Figure 6.20 illustrates this for a relatively soft TPE and a somewhat stiffer PP at a temperature of 40 °C. As the torque increases, the modulus climbs to a maximum value and then falls continuously. Low torques do not ensure optimum force transmission and the specimen slips in the clamp. For this reason, line-shaped pressing are often preferable to the flat type. High torques deform the specimen and lower the moment of resistance, and so the modulus appears to fall. This phenomenon is especially apparent with soft materials. The best torque for clamping each specimen is to be found either at the highest modulus value or in the plateau region.

The specimen must be clamped uniformly and not too tightly.

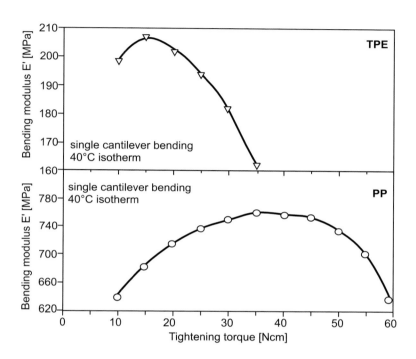

Fig. 6.20   Influence of tightening torque on the elastic modulus of TPE and PP

Temperature 40 °C, bending, frequency 1 Hz

6.2.2.5   Purge Gas

DMA also uses an inert purge gas (nitrogen, helium, or argon) to prevent oxidation. Nitrogen is the most common, for reasons of cost. Helium is recommended when very low starting temperatures have to be reached as quickly as possible.

6.2.2.6   Measuring Program

The starting and finishing temperatures, heating rate, frequency, and deformation must be specified prior to the measurement.

Bulky specimens and heat dissipation into the clamps (the level of which varies from system to system) can create a steep temperature gradient within the specimen that becomes steeper with decrease in starting temperature. It is advisable to start the measurement at 30–50 °C below the expected events, and to hold that temperature constant for several minutes.

Extensive heat dissipation via the clamps
Temperature gradient develops within the specimen.

The finishing temperature is governed primarily by the dimensional stability of the specimen itself. Once the glass transition has been passed, the modulus drops to a low value and the experiment is usually then concluded. When not fullycured thermosets are measured, the drop in storage modulus may be followed by a postcuring reaction and a renewed rise in the modulus (Fig. 5.22); in such cases, a higher finishing temperature should be used.

As the specimens and test chambers are relatively large, the heating rate lies between 1 and 5 °C/min. ISO 6721-1 [1] and ASTM D 4065-2001 [24] recommend a maximum heating rate of 2 °C/min or increments of 2–5 °C, each maintained for 3–5 minutes. DIN 65583 [21] specifies a heating rate of 3 °C/min for fiber-reinforced polymers.

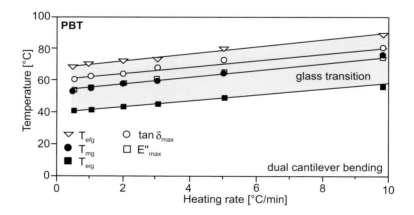

Fig. 6.21   Influence of heating rate on the glass transition temperatures of PBT

T_{efg} = Extrapolated end temperature, $\tan\delta_{max}$ = Peak maximum of the loss factor,
T_{mg} = Midpoint temperature, E''_{max} = Peak maximum of the loss modulus,
T_{eig} = Extrapolated onset temperature,
Dual-cantilever bending, sample geometry 4 x 10 x 14 mm, deformation 10 μm,
frequency 1 Hz

Figure 6.21 shows the effect of the heating rate on the glass transition temperatures of PBT, which increase almost linearly with the heating rate. These temperature differences are due to the increasing difference between the chamber and specimen temperatures as the heating rate increases and vary according to chamber, temperature range, specimen material, and specimen size.

While low heating rates of 0.5 or 1 °C/min ensure uniform thermal equilibration of the specimen and so yield relatively accurate values, they can also greatly affect the specimen's structure (see Section 1.2.2.5). In everyday laboratory work, a balance must be achieved between optimum measuring conditions and a realistic expenditure of time, and that necessarily means reaching a compromise on the heating rate. Measurements are only comparable if conducted at the same heating rate.

The thermocouple should be 3–5 mm away from the specimen.

Best heating rate: 1–3 °C/min

Chemical reactions accompanied by high levels of evolved heat can cause a major, temporary increase in the ambient temperature of the specimen and thus affect temperature control in the apparatus. Fig. 6.22 shows what happened when an incompletely cured vinyl ester (VE) resin specimen was placed too close to the thermocouple. The heat released during postcuring caused the heating rate to fluctuate.

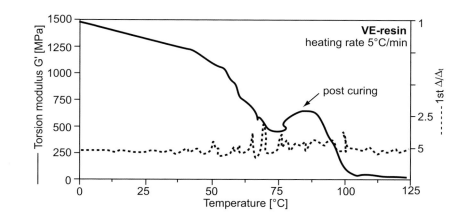

Fig. 6.22   Fluctuations in the heating rate due to postcuring of a VE-resin placed too close to the thermocouple

The right axis shows the heating rate; torsion, heating rate 5 °C/min, frequency 1 Hz

The response of viscoelastic polymers depends on the loading frequency. The higher the frequency, the lower the effect of a nonelastic deformation has – the polymer molecule chains cannot relax in time with the alternating load and the material behaves as if it were stiffer at the same temperature. Higher frequencies shift the glass transition region to higher temperatures, Fig. 6.23.

Events such as melting and crystallization, evaporation, and chemical reactions are not affected by the frequency. Figure 6.24 illustrates this for an SMC paste whose sheer modulus is initially dependent on the frequency, whereas the start of curing is not.

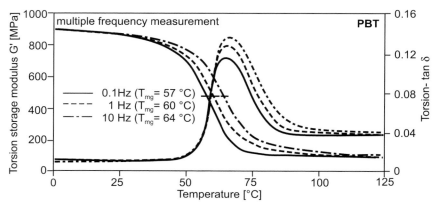

Fig. 6.23   Frequency dependence of storage modulus and loss factor for PBT

T_{mg} = Midpoint temperature, torsion, heating rate 1 °C/min

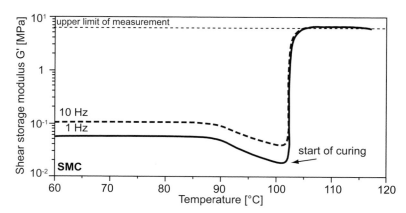

Fig. 6.24   Influence of frequency on modulus and incipient curing of an SMC paste

Shear, heating rate 2 °C/min

Influence of frequency
Increasing the frequency leads to higher T$_g$.
Frequency-independent: Melting, chemical reactions

Deformation is either set as a deflection angle in degrees or radians (360° = 6.28 rad, 1° = 0.01745 rad) for torsional load or as a deformation in micrometers for tensile or flexural load. But this information about the load applied to the material is useful only if stated as percentage deformation for that particular sample geometry (see Section 6.2.2.2). Modern DMA apparatus permits the deformation to be specified as a percentage.

6.2.2.7 T_g-Evaluation
The various methods of determining glass transition temperatures were discussed in Section 6.1.4. But because this area is of huge practical importance and opinions about it differ in practice, we will now examine it in more detail.

Modulus Step

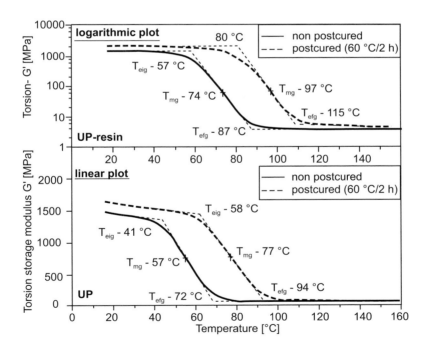

Fig. 6.25   Step evaluation of UP resin specimens, cured (60 °C, 2 h) and uncured

Linear and logarithmic plots, torsion, heating rate 3 °C/min, frequency 1 Hz

In the step method, the extrapolated onset temperature, T_{eig}, the midpoint temperature T_{mg} and the extrapolated end temperature T_{efg} are determined by applying tangents before and after the glass transition.

Figure 6.25 shows how different results are obtained when the storage modulus for UP resin specimens – one uncured and the other cured at 60 °C for 2 hours – is plotted linearly and logarithmically. The T_g values differ by approx. 20 °C for the same degree of cure.

The choice of temperature for the tangents plays a key role in the value of the calculated T_g. The position and slope of the tangents are specified by the operator or the evaluation software in the form of contact points. For steep steps, the tangent is constructed either through the turning point or again by stipulating one or more contact points.

Figure 6.26 shows how the choice of tangent contact points affects the glass transition temperature T_{mg} in the step method.

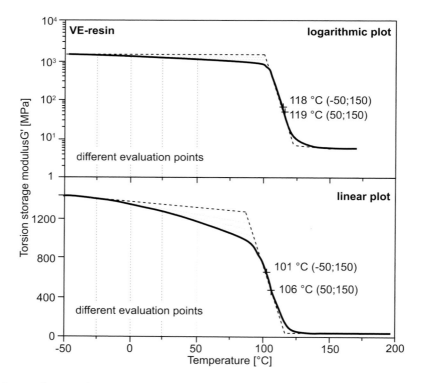

Fig. 6.26   Influence of tangent contact points on the glass transition temperature of a VE resin in the step method; linear and logarithmic plots

Midpoint temperature T_{mg} determined in accordance with Fig. 6.7, torsion, heating rate 5 °C/min, frequency 1 Hz

In a logarithmic plot, T_{mg} varies by only 1 °C as a function of the tangent contact temperature over the range -50 °C to 50 °C. However, the values are 13–17 °C higher than those yielded by the more logical linear plot.

An overview of the corresponding temperatures is provided in Table 6.2 and Table 6.3. Logarithmic plots are found to be less sensitive to differences in the temperatures chosen for the tangents than are linear plots.

1st Tangent [°C]	2nd Tangent [°C]	T_{eig} [°C]	T_{mg} [°C]	T_{efg} [°C]
-50	150	86	101	118
-25	150	91	104	118
0	150	94	106	118
25	150	95	106	118
50	150	95	106	118

Table 6.2   Influence of tangent contact points on glass transition temperatures when storage modulus is plotted linearly

T_{eig} = onset temperature, T_{mg} = midpoint temperature, T_{efg} = end temperature

1st Tangent [°C]	2nd Tangent [°C]	T_{eig} [°C]	T_{mg} [°C]	T_{efg} [°C]
-50	150	105	118	131
-25	150	106	118	131
0	150	106	118	131
25	150	107	119	131
50	150	107	119	131

Table 6.3   Influence of tangent contact points on glass transition temperatures when storage modulus is plotted logarithmically

Step method

Linear plot: Realistic information; evaluation limits have strong influence.
Logarithmic plot: Higher T_g values; evaluation limits have weak influence.

In actual use, components made of high-performance materials are allowed to undergo only minor changes of stiffness. DIN 65583 [21] proposes that the service limits for fiber composites be determined by means of the 2% offset and tangent methods of evaluation.

In the **2% offset method** of DIN 65583 (draft dated 1990), a tangent is applied at 30 °C to a linear plot of the storage modulus and is then offset in parallel to the modulus value by 2%. The intersection of these parallels with the storage modulus curve marks the start of the glass transition, which is called T_{GA}.

Figure 6.27 illustrates this for cured (60 °C, 2 h) and uncured UP resin specimens. It returns T_{GA} temperatures of 36 °C and 46 °C.

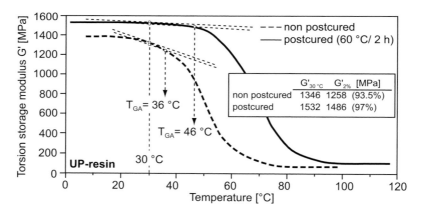

Fig. 6.27   Evaluation by the 2% offset method, DIN 65583 [21], of cured (2 h, 60 °C) and uncured UP-resin specimens, T_{GA} = start of glass transition

Torsion, heating rate 3 °C/min, frequency 1 Hz

The problem with this method is that this new specified reference temperature already lies in the glass transition region (as in the case of the uncured specimen). The modulus drop is much greater as a result -6.5% for the uncured specimen as opposed to 3% for the cure one.

An x% offset in modulus has consequences for practical work.
A suitable reference temperature is important.

Fiber composites may also be evaluated by the **tangent method** set out in DIN 65583. The start of the glass transition, T_W, is the intersection of a tangent applied at 30 °C to a linear plot of storage modulus with the tangent of the turning point of the curve. This roughly corresponds to the extrapolated onset temperature T_{eig} yielded by the step method.

The authors' own experiments have shown that when people manually evaluate "non-ideal" curves using the 2% offset and tangent methods, the subjective nature of manual evaluation yields T_W values that are relatively high – and may be as much as 30 °C too high [36].

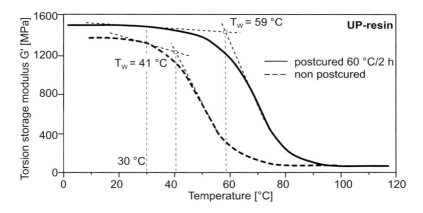

Fig. 6.28   Evaluation by the tangent method, DIN 65583 [21], for cured (2 h at 60 °C) and uncured
               UP resin specimens

               T_W = *Start of glass transition, torsion, heating rate 3 °C/min, frequency 1 Hz*

The above-mentioned weakness in the **draft version** of DIN 65 583 concerning the specified reference temperature of 30 °C was amended with the enactment of DIN 65 583 in April 1999. Evaluation in accordance with the standard is now performed by means of a tangent (T_w becomes T_{g0}) and the 2% method. A reference temperature of $T = T_{g0}$ -50 °C has been introduced for the tangent (T_{GA} becomes $T_{g2\%}$; see also Fig. 6.8).

Peaks
Peaks are much easier to evaluate than tangents.

The **maximum loss factor**, which is frequently interpreted as the glass transition temperature, lies beyond the softening region (Fig. 6.29). For a UP resin specimen, the midpoint temperature T_{mg} (linear; step method) is 23 °C higher than T_g (tan δ_{max}). One factor contributing to the magnitude of this temperature difference is the heating rate.

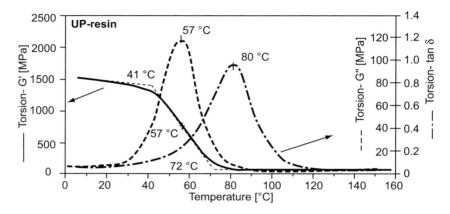

Fig. 6.29   Storage modulus G', loss modulus G'' and loss factor tan δ of an uncured UP resin
              specimen

              Torsion 0.09°, heating rate 3 °C/min, frequency 1 Hz

The maximum loss modulus is frequently used instead of the loss factor. From both physico-chemical and practical points of view, this seems to be an excellent way obtaining a reproducible value for the glass transition temperature that correlates very well in our experience with the T_{mg} value yielded by the step method.

$$T_g\,(\tan \delta_{max}) > T_{mg}, \textbf{ linear}$$
$$T_g\,(G''\text{max}) \approx T_{mg}, \textbf{DMA, linear } \approx T_{mg}, \textbf{DSC}$$

6.2.3 Real-Life Examples

6.2.3.1   DMA Curves for Various Polymers

Polymer technologists, and above all designers, are interested in how plastics respond to heat in practice. Because DMA plots the modulus as a function of temperature while providing high resolution within the glass transition region, it has more to offer in terms of characterization than other thermoanalytical methods. Figure 6.30 is an overview of how the storage moduli of semicrystalline polymers change. In the glass transition region, the amorphous domains soften and the storage modulus drops in steps to a much lower value. Above T_g, the curve starts off level and then tails away when melting starts. The reason that the modulus remains largely constant between the glass transition and the melting regions is the presence of crystalline structures, whose stiffness is only slightly modified by changes in temperature.

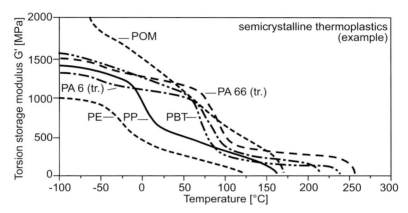

Fig. 6.30   Storage modulus curves for various semicrystalline thermoplastics

Torsion, heating rate 5 °C/min, frequency 1 Hz

By contrast, the storage modulus of amorphous thermoplastics plummets in the glass transition because all the material softens, Fig. 6.31.

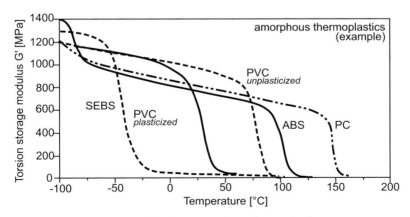

Fig. 6.31   Storage modulus curves of various amorphous thermoplastics

Torsion, heating rate 5 °C/min, 1 Hz

6.2.3.2   Measurements on Polyamide

Figure 6.32 shows the variation in storage modulus, loss modulus, and loss factor of a dry PA specimen over the range -140 to 300 °C, which can be divided into five temperature zones.

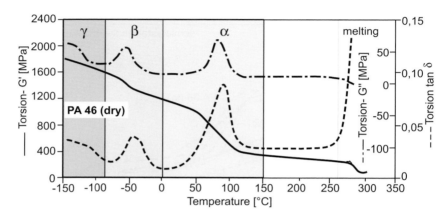

Fig. 6.32   Plot of storage and loss modulus and tanδ for a dry PA 46

Torsion, heating rate 3 °C/min, frequency 1 Hz

There are three so-called relaxation areas, plus a zone of virtual constancy and a zone where the crystalline domains begin to melt. The relaxation zones feature modulus steps and peaks for the tan δ and G'' curves. They are labelled from high to low temperatures with Greek letters (α, β, and γ relaxation). α-Relaxation, which occurs in the highest temperature zone, corresponds to the main glass transition region.

The causes of the events occurring in PA are as follows [37].

Below -80 °C:	Movement of CH_2 groups within the molecule chains
-80 °C to approx. 0 °C:	Movement of chain segments and amide groups
0 °C to approx. 150 °C:	Main glass transition region; segments mobile in amorphous domains
150 °C to 270 °C:	Hardly any changes in properties
Above approx. 270 °C:	Melting of crystalline domains

Figure 6.33 compares various polyamides in the dry state. The top diagram contains logarithmic plots of loss moduli (G'') and the storage modulus G'; the bottom diagram is a linear plot of storage modulus. The modulus at low temperatures and the glass transitions

(relaxations) are not as clear in or are even absent from the logarithmic plot, but incipient melting of the crystalline domains (T > 225 °C) is accentuated.

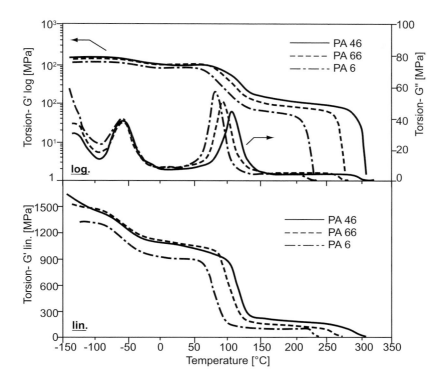

Fig. 6.33   Storage and loss moduli of various dry polyamides of the same geometry

Logarithmic (top) and linear (bottom) plots of storage modulus, torsion
Heating rate 3 °C/min, frequency 1 Hz

The loss modulus curves for the various polyamides show good agreement in the ß-transition region because similar molecular entities become excited at these temperatures. As the intermolecular forces of attraction increase, the main glass transition region, starting with PA 6, through PA 66 to PA 46, is shifted to higher temperatures.

The decrease in magnitude of the loss modulus (and tan δ), which occurs in the same sequence, indicates that the specimens differ in crystallinity [37]. PA 6 is less crystalline than PA 66, which is less crystalline than PA 46.

6.2.3.3   Conditioning – Water Content

The water content of polyamides, especially greatly affects the shape of their DMA curves and the position of the glass transition temperature (see Sections 1.2.3.4 and 4.2.3.2).

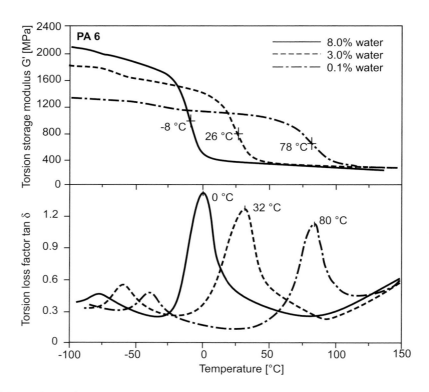

Fig. 6.34   Change in storage modulus and loss factor for PA 6 specimens of different water content

Torsion, heating rate 3 °C/min, frequency 1 Hz

The water inside the specimens increases segment mobility to such an extent that the water could be said to have a plasticizing effect. Figure 6.34 shows the effects of different conditioning states (0.1%, 3.0% and 8.0% H_2O) on the change in storage modulus and loss factor of a PA 6. The midpoint temperature and tan δ_{max} have shifted by about 80 °C. Below 0 °C, the occluded water causes the stiffness to rise on account of so-called freezing effects [29, 38, 39].

Figure 6.35 shows how the water content of pure EP resin specimens affects their DMA curves. The modulus of the dry specimen is initially constant. Although it decreases steadily in the moist specimen, it is starting from a higher value because of the aforementioned freezing effect. From approx. 100 °C, the modulus drops as a sharp step, and this may be attributed to the release of water (desorption). This effect is superimposed on the glass transition, which appears to start at approx. 125 °C, but this cannot be established with certainty.

The modulus curve of the dry resin, by contrast, has an almost ideal shape, and its glass transition lies at much higher temperatures.

Tan δ curves make it relatively easy to determine the glass transition; their peaks are very close to each other. However, the curves differ considerably in shape.

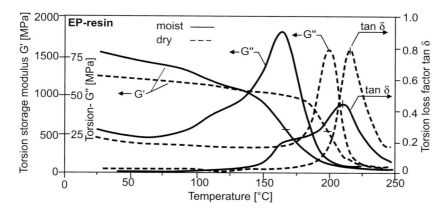

Fig. 6.35   Change in storage modulus, loss modulus, and loss factor of EP resin specimens
               conditioned to different extents

               Torsion, heating rate 3 °C/min, frequency 1 Hz

Table 6.4 shows the T_g values as determined from the modulus curves in Fig. 6.35.

	T_{eig}	T_{mg}	$T_g (\tan \delta_{max})$	$T_g (G''_{max})$
Moist	140 °C	162°C/168 °C*	210 °C	162 °C
Dry	186 °C	200 °C	215 °C	200 °C

               * *Different evaluation limits*

Table 6.4   Different T_g values for the EP resin specimens (wet/dry) in Fig. 6.35

The differences are found to be 46 °C for T_{eig} (extrapolated onset temperature), approx. 40 °C for T_{mg} (midpoint temperature) and just 5 °C for T_g (tan δ_{max}). The maximum loss modulus is like T_{mg}, but can be determined more readily than the midpoint temperature and should be preferred in such a case. Designers, however, should note the temperature at which each material softens.

Moisture affects the modulus and shifts T_g to lower temperatures.

6.2.3.4   Blends

DMA is ideal for analyzing blends in the glass transition as it provides very good resolution in this region.

Compatible blends

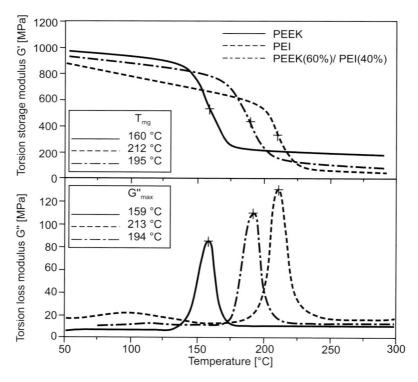

Fig. 6.36   Storage and loss moduli of PEEK, PEI, and a compatible blend of both components
PEEK/PEI (60/40)

Torsion, heating rate 5 °C/min, frequency 1 Hz

Figure 6.36 shows the storage and loss modulus curves for a blend and its semicrystalline PEEK and amorphous PEI constituents. Because this is a compatible blend, it exhibits a single glass transition located between those of the two pure components (see Section 1.2.3.8). The level of torsional stiffness above the glass transition region varies with the fraction of crystalline domains; there are more of these in the PEEK than in the blend.

Incompatible blends

By contrast, each constituent of an incompatible blend exhibits its own characteristic glass transition. Figure 6.37 shows the storage and loss moduli of a blend composed of semi-crystalline PEKEKK and amorphous PES along with those for the individual constituents.

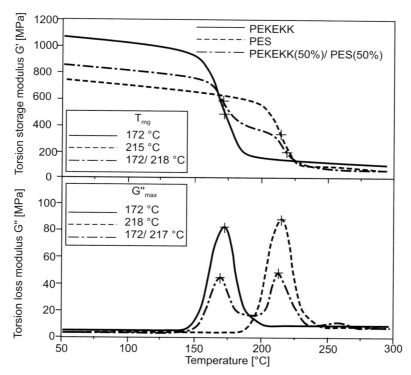

Fig. 6.37   Storage and loss moduli of PEKEKK, PES, and a compatible blend of both components PEKEKK/PES (50/50)

Torsion, heating rate 5 °C/min, frequency 1 Hz

Compatible blends: Common glass transition
Incompatible blends: Two glass transitions

DMA can also be used to quantify the proportions of the components in the blend – via the relative height of the peak maximum or the peak area of the loss modulus.

This is illustrated in Fig. 6.38, which shows the loss moduli for blends of different compositions. The relative (unknown) proportions of components in a blend can be quantified by reference to a calibration curve (see also Section 1.2.3.8).

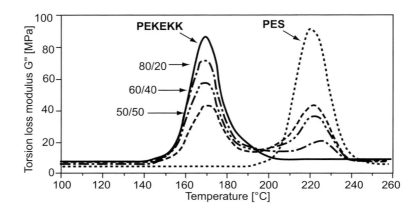

Fig. 6.38   Loss modulus curves for various incompatible blends of PEKEKK and PES

Torsion, heating rate 5 °C/min, frequency 1 Hz

As already outlined in Section 1.2.3.8 (DSC), the dry impact strength of polyamides may be improved by modification with polyethylene or elastomers. The polyethylene can be rapidly identified by DSC, but elastomers, which are usually present only in small amounts, are more readily characterized by DMA.

Figure 6.39 shows the change in storage and loss moduli of pure and elastomer-modified PA 66. The step in the storage modulus at low temperatures indicates the presence of an elastomeric component; however, this is also the region of ß-relaxation. The loss modulus curve reveals both events in the form of two adjacent peaks at approx. -72 °C (ß-relaxation) and -45 °C (elastomer).

Loss factor curves show equally good resolution (not shown).

The shape of the loss factor and loss modulus curves also reveals the presence of a blend.

Polymer modification is identifiable from loss modulus and loss factor.

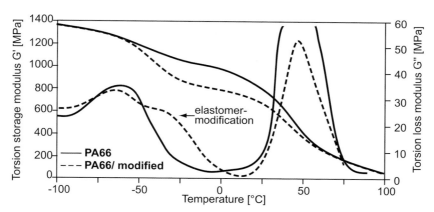

Fig. 6.39   Storage and loss modulus of pure and elastomer-modified PA 66

Torsion, heating rate 5 °C/min, frequency 1 Hz

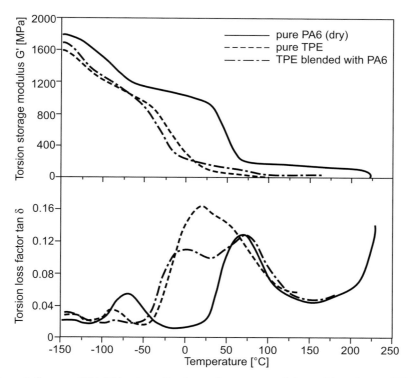

Fig. 6.40   Influence of PA 6 blends on the change in storage modulus and loss factor of TPE

Torsion, heating rate 5 °C/min, frequency 1 Hz

The upper part of Fig. 6.40 shows the storage modulus curves for pure PA 6, TPE and a TPE probably blended with PA during drying or filling of the injection molding machine; the corresponding loss modulus curves are shown in the lower part. The storage modulus curves only hint at the presence of a blend but do not provide definitive information. However, the loss modulus curve for the suspected blend contains two peaks in the glass transitions of the two materials, and this indicates that a blend is present.

6.2.3.5 Annealing

As described in Sections 1.2.3.3 and 4.2.3.1, annealing semicrystalline materials increases their crystallinity. This increase manifests itself in the modulus.

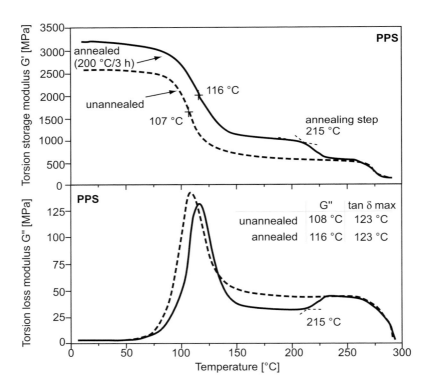

Fig. 6.41   Storage and loss moduli of annealed (3 h, 200 °C) and unannealed PPS
              Loss factor is not shown for reasons of clarity

Torsion, heating rate 3 °C/min, frequency 1 Hz

Figure 6.41 presents the curves for an annealed (3 h, 200 °C) and untreated PPS specimen. The glass transition temperature T_{mg} of the stiffer specimen is almost 10 °C higher. The loss modulus curve (lower diagram) also reveals that the annealed material has a higher glass transition temperature. By contrast, the tan δ_{max} values are the same (see inset table). Above the annealing temperature, the modulus drops back to the level of the unannealed specimen as the less stable crystallites formed at the annealing temperature start melting.

Postcuring of semicrystalline thermoplastics may be identified from the following:

– Increased stiffness up to annealing temperature
– T_g shift to higher temperatures
– Less damping up to annealing temperature

6.2.3.6 Curing of Reaction Resins

As previously shown in the discussions of DSC and TMA (Sections 1.2.3.9 and 4.2.3.4), the glass transition can be used to gage the degree of cure in a thermoset. Figure 6.42 illustrates the shift in glass transition for a VE resin cured at room temperature for 25 h and then postcured at 80 °C for different periods (0.5 h, 1 h, 4 h). Progressive crosslinking with length of postcuring shows up as an increase in glass transition temperature, identifiable from G''_{max} and the midpoint temperature on the G' curve. Furthermore, postcuring increases the storage modulus in the temperature range up to T_g, and the peak maximum of the loss modulus lies at lower absolute values (similar considerations apply to the loss factor, which is not shown here).

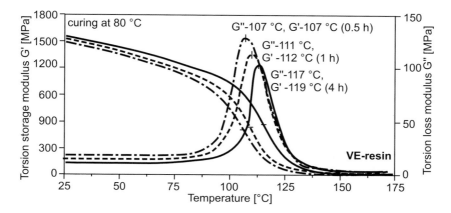

Fig. 6.42   Influence of postcuring on storage and loss moduli of VE-resin cured at 80 °C

Torsion, heating rate 3 °C/min, frequency 1 Hz

Degree of cure can be determined accurately from glass transition temperature.

Incompletely cured specimens sometimes undergo postcuring during the heat-up phase in the apparatus. The postcuring shows up as a rise in storage modulus and as twin peaks in the loss modulus or loss factor. This is illustrated in Fig. 6.43 for VE resin specimens cured at 23 °C (initial cure) and then postcured at 60 °C for different periods (1 and 8 weeks).

All specimens display signs of incipient postcuring in the upper section of the glass transition region, with the greatest increase in storage modulus exhibited by those specimens with the least initial cure (upper diagram); similarly, two peaks occur in the loss modulus, the second of which can vary in size (lower diagram).

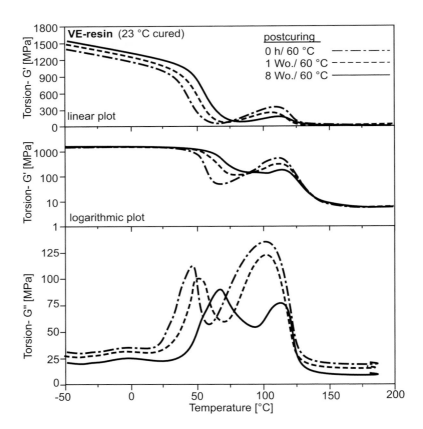

Fig. 6.43   Effect of postcuring on storage and loss moduli of a VE-resin cured for
                 different periods at 23 °C

                 Top: Llinear plot of storage modulus G′
                 Center:Llogarithmic plot of storage modulus G′
                 Bottom: Llinear plot of loss modulus G′′
                 Torsion, heating rate 3 °C/min, frequency 1 Hz

Note: *The linear plot reveals the differences in storage moduli below T_g, while the logarithmic plot reveals those above T_g.*

Postcuring reactions are identifiable from a rise in modulus and a second peak in the loss modulus or loss factor curves.

Figure 6.44 shows the postcuring effects in a high-performance EP resin cast in a film thickness of 50 μm. During the first heating phase (1st heating scan), the postcuring manifests itself as a rise in storage modulus above the softening range and as twin peaks in the loss modulus. During the second heating phase (2nd heating scan), the adhesive is fully cured, much stiffer (higher storage modulus), and softens 60 °C higher.

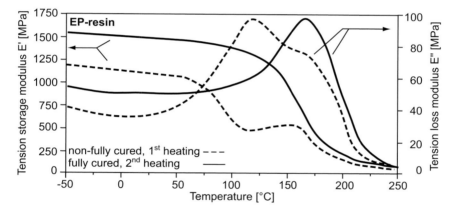

Fig. 6.44   Change in E′ and E′′ of a not fully cured (1st run) and a fully cured adhesive film (2nd run) based on EP resin

Tension, heating rate 2 °C/min, frequency 1 Hz

6.2.3.7   Aging

Thermal and thermooxidative aging of polymers can alter the position of the glass transition of a polymer. Exposing elastomers to repeated processing or high service temperatures can cause adjacent molecule chains to crosslink, and increase and broaden the glass transition [40]. Figure 6.45 shows the influence of repeated injection molding (one to five times) at 260 °C on ABS. Elevated temperatures or mechanical stress or both can rupture double bonds on the butadiene rubber and cause crosslinking. This manifests itself as rising T_{mg} values and falling tan δ values for the butadiene glass transition.

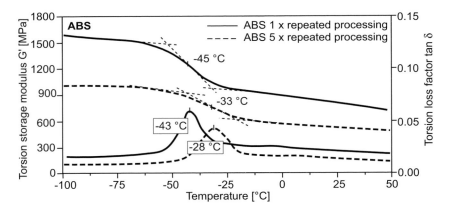

Fig. 6.45   Influence of repeated processing on G′ and tan δ of ABS

Torsion, heating rate 3 °C/min, frequency 1 Hz

Elastomers: Age by crosslinking – T_g rises.

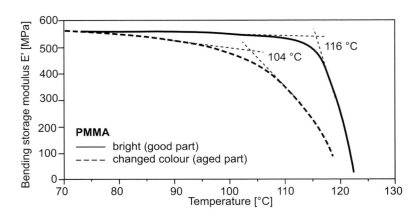

Fig. 6.46   Influence of thermal aging on the storage modulus of PMMA, T_{eig} = extrapolated onset temperature

Cantilever bending, heating rate 2 °C/min, frequency 1 Hz

Amorphous thermoplastics, on the other hand, tend to experience a drop in T_g when aging takes the form of molecular chain degradation (see Section 1.2.3.6). Figure 6.46 compares two molded PMMA parts: a "good part" which was unstressed and a "bad part" which was exposed to elevated service temperatures and UV light. OIT revealed that the two parts had different levels of stabilization. Thermooxidative attack of the inadequately stabilized part ruptured and shortened the molecular chains. Figure 6.46 shows only the first part of the glass transition because in this case the onset temperature seemed to be the most appropriate for the determination. The difference turned out to be approx. 8 °C.

Thermoplastics: Age by chain scission – T_g falls.

6.2.3.8   Influence of Plasticizers

The stiffness of polymers van be lowered to the desired service range by adding plasticizers. PVC (rigid and plasticized) is a good example of this. Figure 6.47 illustrates how the storage modulus can be used to distinguish between PVC grades containing different levels of plasticizer. The higher the plasticizer content, the lower is the glass transition temperature T_{eig} or T_{mg}.

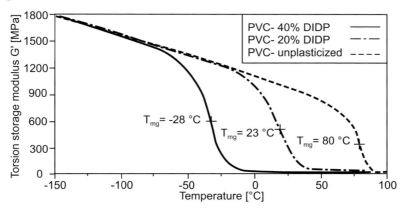

Fig. 6.47   Storage modulus of PVC specimens containing different levels of plasticizer;
              DIDP = Diisodecyl phthalate [41]

              T_{mg} = Midpoint temperature

6.2.3.9   Temperature Distribution in Fiber-Reinforced Polymers

As explained in Section 6.1.5.1, temperature control poses a particular problem in DMA because the specimen chamber is large and the specimens are relatively bulky.

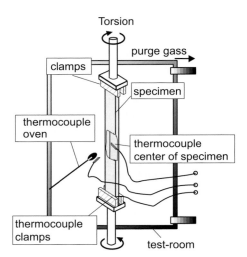

Fig. 6.48   Positioning thermocouples to check the temperature control of
                 DMA measuring systems [29]

Discrepancies in the temperature measured by a thermocouple close to the specimen and the
specimen's actual temperature are due to the heat incorporated into the specimen as a
function of heating rate and specimen geometry as well as to dissipation of heat via the
clamps, which are usually metallic.

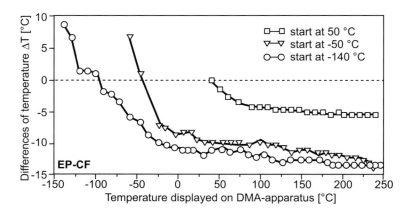

Fig. 6.49   Differences between temperature displayed on DMA apparatus and the temperature
                 measured at the center of the specimen as a function of starting temperature (quasi-
                 isotropic EP-CF specimen [29]

                 Torsion, heating rate 3 °C/min, frequency 1Hz

A round-robin test was conducted on heat-curing EP-CF resin to test the temperature control systems of different DMA apparatues. To this end, the centers of the 10 mm wide specimens were drilled to a depth of 5 mm (∅ 0.7 mm). Thermocouples (type K, wire ∅ < 0.25 mm) were bonded inside these holes with an epoxy resin adhesive that exhibited no thermal effects in the temperature range of interest [29, 36]. This arrangement is shown in Fig. 6.48. The temperature change in the specimens was recorded during the experiments and compared with the set values on the apparatus. To log temperature gradients along each specimen, two further thermocouples were mounted both at the clamps and between the clamps and the center of the specimen. The difference between the temperature displayed on each apparatus and the specimen's actual temperature (at its center), ΔT, grew larger and larger Fig. 6.49 and was especially pronounced at very low starting temperatures.

The observed differences vary from 5 to approx. 14 °C according to starting temperature and temperature range. This means that the specimen's temperature lags behind by about 5 °C when the starting temperature is 50 °C. At very low temperatures, the specimen initially remains warmer than the environment for a period but, from -100 °C, the situation changes and the specimen is much colder than its environment. This latter effect depends greatly on heat conduction in the specimen. Especially when thermal conduction is anisotropic the case of fiber-reinforced polymers in, extensive heat conduction may occur if the fibers are aligned with the clamps. This would be especially the case for high-modulus carbon fibers with a thermal conductivity of λ = 115 W/(m K) as opposed to glass-fibers (λ = 1 W/[m K]) and aramid fibers (λ = 0.05 W/[m K]), (cf. aluminum λ = 200 W/[m K]), iron λ = 81 W/[m K]) [29].

Starting temperature and thermal conductivity of the reinforcing material influence the specimen's temperature.

Because some types of apparatus offer the possibility of a two-point temperature calibration, the specimen's actual temperature should always be checked with the aid of integrated thermocouples where there is any doubt.

Figure 6.50 shows the influence of CRP fiber orientation on the difference in temperature displayed by the apparatus and the specimen's actual temperature (measured at the center) for ± 45° and 0° laminates. The actual temperature of the ± 45° specimen over the range 0–150 °C is 5–7 °C lower than that displayed while that of the unidirectional specimen is as much as 10–17 °C lower. The cause of the latter effect is that the heat is transferred directly through the highly conductive carbon fibers arranged in parallel between the clamps, which themselves are still very cold in this temperature range.

Fiber orientation influences the temperature distribution in the specimen.

Furthermore, the temperature was measured at various points on the specimens (not shown in the diagram). A considerable temperature gradient was observed along the unclamped section. These problems mean that it is doubtful whether a specimen's temperature and thus its glass transition temperature can be correctly quoted.

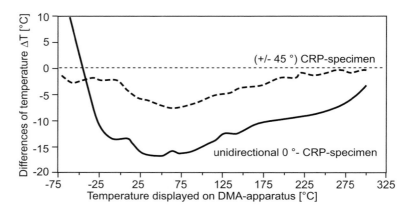

Fig. 6.50   Differences between temperature displayed on DMA apparatus and the temperature measured at the center of the specimen (± 45° and 0°); CRP specimen [29]

Torsion, heating rate 3 °C/min, frequency 1 Hz

6.3   References

[1]   N.N.   ISO 6721-1 (2001)
                                     Plastics – Determination of Dynamic Mechanical
                                     Properties – Part 1: General Principles

[2]   N.N.   ASTM D 4092 (2001)
                                     Standard Terminology – Plastics –
                                     Dynamic Mechanical Properties

[3]   Münstedt, H.   Rheology – Applications in and Relevance to Industrial
                                     Practice, Proceedings
                                     Lehrstuhl für Polymerwerkstoffe, Erlangen 1997

[4]   Schwarzl, F.R.   Polymermechanik - Struktur and mechanisches
                              Verhalten von Polymeren
                              Springer-Verlag, Berlin 1990

[5]   Gallagher, P.K.,   Dynamic Mechanical Analysis (DMA)
      Turi, E.A.   Thermal Characterization of Polymeric Materials,
                              Second Edition
                              Academic Press, San Diego 1997

[6]   N.N.   ASTM D 5279 (2001)
                              Standard Test Method for Plastics: Dynamic Mechanical
                              Properties: Torsion

[7]   N.N.   ISO 6721-7 (2003) DAM 1
                              Plastics – Determination of Dynamic Mechanical
                              Properties – Part 7: Torsional Vibration –
                              Non-Resonance Method / Note: Intended as an
                              Amendment to ISO 6721-7 (1996)

[8]   N.N.   ISO 6721-2 (1995)
                              Plastics – Determination of Dynamic Mechanical
                              Properties – Part 2: Torion-Pendulum Method

[9]   N.N.   ISO 6721-3 (1995)
                              Plastics – Determination of Dynamic Mechanical
                              Properties – Part 3: Flexural Vibration – Resonance-
                              Curve Method / Note: Technical Corrigendum 1

[10] N.N.   Polymer Laboratories
                              Operator's Guide DMTA MkII

[11] N.N.   ISO 6721-10 (1999)
                              Plastics – Determination of Dynamic Mechanical
                              Properties – Part 10: Complex Shear Viscosity Using a
                              Parallel-Plate Oscillatory Rheometer

[12] N.N.   ASTM D 4440 (2001)
                              Standard Test Method for Plastics –
                              Dynamic Mechanical Properties – Melt Rheology

[13]   N.N.   ASTM D 5023 (2001)
                                 Standard Test Method for Measuring the
                                 Dynamic Mechanical Properties –
                                 In Flexure (Three-Point-Bending)

[14]   N.N.   ISO 6721-5 (2003) DAM 1
                                 Plastics – Determination of Dynamic Mechanical
                                 Properties – Part 5: Flexural Vibration – Non-
                                 Resonance Method / Note: Intended as an Amendment
                                 to ISO 6721-5 (1996)

[15]   N.N.   ASTM D 5418 (2001)
                                 Standard Test Method for Plastics –
                                 Dynamic Mechanical Properties – In Flexure
                                 (Dual Cantilever Beam)

[16]   N.N.   ASTM D 5026 (2001)
                                 Standard Test Method for Plastics –
                                 Dynamic Mechanical Properties – In Tension

[17]   N.N.   ISO 6721-4 (1994)
                                 Plastics – Determination of Dynamic Mechanical
                                 Properties – Part 4: Tensile Vibration – Non-Resonance-
                                 Method

[18]   N.N.   ASTM D 5024 (2001)
                                 Standard Test Method for Plastics – Dynamic
                                 Mechanical Properties – In Compression

[19]   N.N.   ISO 6721-6 (2003) DAM 1
                                 Plastics – Determination of Dynamic Mechanical
                                 Properties – Part 6: Shear Vibration –
                                 Non-Resonance Method / Note: Intended as an
                                 Amendment to ISO 6721–6 (1996)

[20]   Chartoff, R.P.,   Thermoplastic Polymers
       Turi, E.   Thermal Characterization of Polymeric Materials,
                                 Second Edition
                                 Academic Press, San Diego 1997

[21]   N.N.   DIN 65 583 (1999)
                                            Aerospace – Fibre Reinforced Materials – Determination
                                            of Glass Transition of Fibre Composites Under
                                            Dynamic Load

[22]   N.N.   ISO 11357-1 (1997)
                                            Plastics – Differential Scanning Calorimetry (DSC) –
                                            Part 1: General Principles

[23]   Rieger, J.   The Glass Temperature T_g of Polymers –
                                            Comparison of the Values yielded by DSC and DMA
                                            Measurements (Torsion Pendulum)
                                            Personal Information, February 1998

[24]   N.N.   ASTM D 4065 (2001)
                                            Standard Practice for Plastics – Dynamic Mechanical
                                            Properties – Determining and Reporting of Procedures

[25]   N.N.   DIN 53 765 (1994)
                                            Testing of Plastics an Elastomers – Thermal Analysis;
                                            DSC-Method

[26]   Ferrillo, R.G.,   Comparison of Thermal Techniques for Glass Transition
         Achorn, P.J.   Assignment. II. Commercial Polymers
                                            Journal of Applied Polymer Science 64 (1997), p. 191

[27]   N.N.   DIN 53 545 (1990)
                                            Determination of Low Temperature Behaviour of
                                            Elastomers – Principles and Test Methods

[28]   N.N.   ASTM E 1640 (1999)
                                            Standard Test Method for Assignment of the Glass
                                            Transition Temperature by Dynamic Mechanical
                                            Analysis

[29]   Schemme, M.,   Charakterisierung von Faserverbund-Kunststoffen
         Avondet, M.A.,   mit Methoden der Dynamisch-Mechanischen Analyse
         Ehrenstein, G.W.   (DMA)
                                            Materialprüfung 39 (1997) 3, pp. 59–66

[30]   N.N.   ASTM E 1867 (2001)
                                            Standard Test Method for Temperature Calibration
                                            of Dynamic Mechanical Analyzers

[31] N.N.   ASTM E 2254 (2003)
                                  Standard Test Method for Storage Modulus Calibration
                                  of Dynamic Mechanical Analyzers

[32] Ehrenstein, G.W.   Mit Kunststoffen konstruieren, Second Edition
                                  Carl Hanser Publishers, Munich 2001

[33] Retting, W.,   Kunststoff-Physik
      Laun, H.-M.   Carl Hanser Publishers, Munich 1991

[34] Pahl, M., Gleißle, W.,   Praktische Rheologie der Kunststoffe and Elastomere
      Laun, H.-M.   VDI -Verlag GmbH, Düsseldorf 1995

[35] Beitz, W.,   Dubbel - Taschenbuch für den Maschinenbau
      Küttner, K.H.   Springer Verlag, Berlin 1987, 16, C29

[36] Wolfrum, J.,   Dynamic Mecanical Analysis of High-performance
      Ehrenstein, G.W.,   Composites – Influences and Problems
      Avondet, M.A.   Lecture on Composite Materials and Material
                                  Composites, 17.-19.9.1997 in Kaiserslautern
                                  DGM Informationsgesellschaft mbH 1997

[37] Schmack, G.,   Strukturbeeinflussung von Polyamiden
      Vogel, R.,   Kunststoffe 84 (1994) 11, pp. 1590–1594
      Häußler, L.

[38] Baschek, G.,   Effect of Water Absorption in Polymers at Low and
      Hartwig, G.,   High Temperatures
      Zahradnik, F.   Polymer 40 (1999), pp. 3433–3441

[39] Birkinshaw, C.,   Plasticization of Nylon 66 by Water and Alcohols
      Buggy, M.,   Polymer Communications, 28 (1987), pp. 286–288
      Daly, S.

[40] Weidner, H.,   Alterungsvorhersagen bei Kunststoffen -
      Tilger, R.   Charakterisierung von Alterungszuständen
                                  Kunststoffe 70 (1980) 12, pp. 837–844

[41] Ehrenstein, G.W.   Polymeric Materials
                                  Carl Hanser Publishers, Munich 2001

Other References:

[42]   N.N.   ISO 6721-8 (1997)
                                   Plastics – Determination of Dynamic Mechanical
                                   Properties – Part 8: Longitudinal and Shear Vibration –
                                   Wave-Propagation Method

[43]   N.N.   ISO 6721-9 (1997)
                                   Plastics – Determination of Dynamic Mechanical
                                   Properties – Part 9: Tensile Vibration – Sonic-Pulse
                                   Propagation Method

[44]   N.N.   ASTM D 4473 (2003)
                                   Standard Test Method for Plastics –
                                   Dynamic Mechanical Properties – Cure Behavior

[45]   N.N.   ASTM D 6048 (2002)
                                   Standard Practice for Stress Relaxation Testing of Raw
                                   Rubber, Unvulcanized Rubber Compounds and
                                   Thermoplastic Elastomers

[46]   Wingfield, M.   Bestimmung des Glasübergangs per DMTA
                                   LaborPraxis, September 1996, pp. 76–89

[47]   Möhler, H.   Die Thermische Analyse in der Kunststoffprüfung
                                   Kunststoffe 84 (1994) 6, pp. 736–743

[48]   N.N.   Rheometric Scientific
                                   Operator's Guide DMTA MkIV
                                   Rheometric Scientific 1996

[49]   Matsuoka, S.   Viscoelastic Properties of Polymers
                                   Seminarband, Polytechnic University Brooklyn,
                                   New York 1997

[50]   Ward, I. M.   Mechanical Properties of Solid Polymers,
                                   Second Edition
                                   John Wiley & Sons, Chichester 1983

[51]   Williams, J.G.   Fracture Mechanics of Polymers
                                   John Wiley & Sons, Chichester 1987

7   Micro-Thermal Analysis

7.1   Principles of micro-TA

7.1.1 Introduction

Micro-thermal analysis (μTA) is a combination of thermal analysis and atomic force microscopy. The instrument is basically an atomic force microscope (AFM) fitted with a tip that acts as temperature sensor and heat source. This configuration permits thermal characterization of subsurface regions in addition to the localized high-resolution topography-imaging ability of conventional AFM.

> **Micro-TA – Combination of atomic force microscopy (AFM)**
> **and thermal analysis (TA)**

7.1.2 Measuring Principle

In atomic force microscopy, a sharp tip mounted on the end of a cantilever is swept over the surface of the sample in a raster scan. Variations in topography cause the tip and cantilever to deflect. A mirror attached to the cantilever reflects a laser beam onto a split photodiode. A feedback loop controls the position of the cantilever mount to keep the bending of the cantilever constant. A plot of the vertical movement of the cantilever mount as a function of the position of the tip in the raster yields a quantitative, three-dimensional image of the topography.

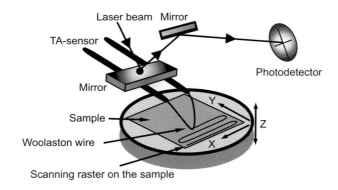

Fig. 7.1   Schematic diagram of the μTA™ [1]

In micro-TA, the conventional tip of the AFM is exchanged for a miniature heating filament that simultaneously acts as a temperature sensor. The tip is made from Wollaston wire consisting of a silver wire of 50 μm diameter containing a platinum core of just 5 μm diameter. A short section of the silver is removed to expose the platinum core, which is bent into the shape of the tip, Fig. 7.1.

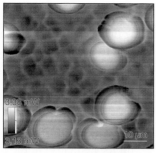

Thermal conductivity image

Glass fibers
(high thermal conductivity)

Spherolites
(bright edges caused by currature at the spherolite edge and thus changes in contact area between the sensor and the sample)

Topography image

Polishing the sample enhances emphasizes the soft polymer matrix and especially the amorphous components. The contours of glass fibers and spherolites are rendered visible. If the differences in topography are too large, poor contact leads to artefacts in the thermal conductivity image.

Shaded 3D topography

The image can be shown in 3D for better presentation and contour definition.

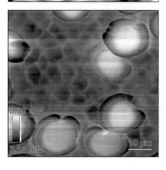

Fig. 7.2   Scans of a glass-fiber-reinforced PA 66 sample (polished surface)

        Top: Thermal conductivity differences,
        Center: Topography
        Bottom: Shaded 3D topography

The jacketed wire ends act as a cantilever. An attached mirror for reflecting the laser beam affords a means of imaging the topography. When an electric current flows through the platinum wire, the wire heats up at the point of maximum ohmic resistance, that is, at its tip. If the tip is in direct contact with the sample, as it is during rastering, its temperature drops due to heat transfer to the specimen. The electrical power needed to keep the tip at a predetermined temperature is then monitored. This information can be used to produce an image of the local variation in thermal conductivity near the surface of the sample.

Figure 7.2 shows typical topography and thermal conductivity images for a glass-fiber-rein-forced PA 66. The thermal conductivity differences between matrix and glass fibers may be clearly seen. Topography can be visualized more readily by shading (bottom). Once these images have been obtained, further analyses can be conducted by subjecting localized, high-resolution areas to micro-DTA and micro-TMA. These methods involve the use of a temperature program to heat the tip.

The micro-DTA experiments measure the electrical power needed to keep the heating rate constant. This is done with the aid of an identical reference tip located close to the point of measurement. Micro-TMA curves (of penetration) recorded concurrently with micro-DTA show the change in vertical deflection of the tip as a function of temperature [2].

Note: *The measurements cannot include the specimen size (weight). These are therefore*
         μDTA determinations and not μDSC determinations.

7.1.3 Procedure and Influencing Factors

Factors affecting the instruments and the samples are:

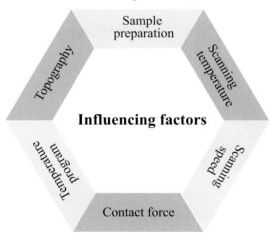

The steps involved in a micro-TA determination are:
 – Prepare the sample.

– Adjust, thoroughly heat and calibrate the sensor.
– Position the plane-parallel sample under the sensor.
– Place the sensor on the area of interest.
– Obtain a thermal conductivity and/or topography image, as required.
– Select measuring points for thermal analysis (TA).
– Set a suitable measuring program (temperature program, frequency, load).

The influencing factors and possible errors that may arise during the procedure are treated in more detail in Section 7.2.2 with the aid of experimental curves.

7.1.4 Evaluation

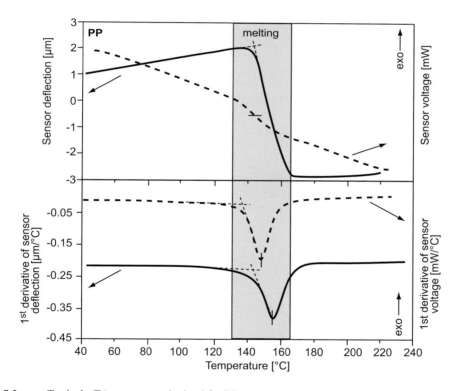

Fig. 7.3   Typical µTA curves, as obtained for PP

Top: Sensor deflection and sensor voltage
Bottom: First derivative of the two signals
Heating rate 10 °C/s, load 30 nA

Like DSC, micro-TA can detect endothermic and exothermic effects caused by rises and falls in the enthalpy of the material. Unlike DSC, it cannot determine the underlying sample weight, and so enthalpies cannot be quantified. Evaluation is based on the change in sensor voltage as a function of temperature or time or both. As in TMA, it is also possible to record the deflection of the sensor in micrometers. Deflection may be due to expansion of the sample or to penetration of the sample by the sensor (e.g., during melting or at the glass transition).

Figure 7.3 shows the two typical curves, as obtained for PP. The temperatures are determined by means of the tangent intersections. The first derivative of the curve of sensor signal against temperature is also plotted because it makes evaluation easier. The gray area indicates the melting range of polypropylene. The melting point may now be determined either from the intersection of the tangents (onset temperature) applied to the deflection signal or from the half-step height of the sensor voltage or as the onset temperature (tangent intersection) of the first derivative of the sensor voltage. The sensor-deflection signal is used more often than the sensor voltage because it is pronounced in a large number of polymers.

7.1.5 Calibration

Unlike classic DSC and TMA, micro-TA requires the temperature to be calibrated before each new sample is measured.

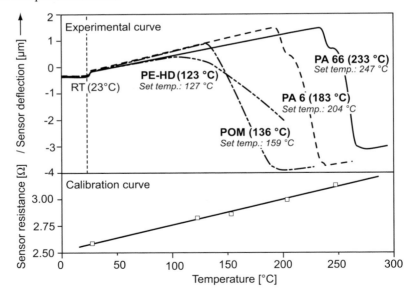

Fig. 7.4   µTA measurements on thermoplastics for compiling the calibration curve

Heating rate 10 °C/s, load 30 nA,
Set temperatures are extrapolated onset temperatures determined by means of DSC

If the AFM is altered or moved in any way, it has to be recalibrated. Calibration is carried out at room temperature with the aid of known substances in the temperature range of interest, Fig. 7.4.

Experience shows that at least three calibrating substances should be used because fluctuations are common.

The I/O-T offset must be calibrated in addition to the temperature. This corrects hysteresis effects within the temperature controller. The longer the controller is left switched on, the more these effects diminish; they should therefore be left on, if possible. It has also been found that the instrument and the measuring electronics are highly sensitive to ambient changes. The determinations must therefore be performed in a permanently air-conditioned environment. Furthermore, calibration must be performed several times a day.

> **A permanently air-conditioned and vibration-free environment is necessary.**
> **Calibration several times a day**

7.1.6   Overview of Practical Applications

Application	Characteristic	Example
Multilayer systems	T_g, T_m	Identifying the individual layers of multi-layer pipes or films
Bonds and joints	T_g, T_m	Characterizing the thickness and curing of coatings
Multicomponent injection molding	T_g, T_m	Analysis of interfacial layers and mixed phases
Aging phenomena	T_m	Characterizing boundary layers

Table 7.1   Examples of practical applications of µTA measurements on polymers

7.2   Procedure

7.2.1 In a Nutshell

Sample preparation

The point of sampling is chosen according to the problem. The sample surface must be smooth (10 µm), polished if possible, and plane-parallel.

Sample positioning

The maximum scanning area for imaging purposes is 100 x 100 µm. The points for localized thermal analysis are chosen after the surface scan.

> **Plane-parallel and smooth (max. 10 µm) samples**

Sensor preparation

The sensor must be cleaned before the determination. Clinging sample residues lead to spurious results. The sensor tip is applied under a suitable pressure (contact force) to the sample surface. The resultant deflection of the laser beam must then be monitored. The quality of the results also depends on the shape and the angle of the tip.

Application of load (contact force)

The load on the sensor tip is switched to nA, and for polymer corresponds to a deflection of between 10 and 50 nA on the photo detector.

Scanning parameters

The temperature of the sensor tip during scanning is constant, usually around 50 °C. It may be increased to highlight certain effects. The scan yields information about the topography of the sample and qualitative differences in thermal conductivity. The images generated therefrom are used to select specific points for localized thermal analysis.

Measuring program

The measuring program must be chosen such that the starting/end temperatures and the heating rate are appropriate for the sample and its constituent material.

Heating rate

The heating rates are very high in comparison with the other methods of thermal analysis. Because the sample has a small surface area and the sensor has a small diameter (5 µm), heating rates of up to 100 °C/s are feasible. Heating rates of 5 to 10 °C/s are suitable for polymers.

Starting temperature

The starting temperature should be at least 50 °C below the first expected effect. Measurements mostly begin at room temperature. Lower starting temperatures are also possible, however. These require an accessory that helps to cool the whole sample.

Heating rates of 5–10 °C/s

Starting temperature

The starting temperature should be at least 50 °C below the first expected effect. Measurements mostly begin at room temperature. Lower starting temperatures are also possible, however. These require an accessory that helps to cool the whole sample.

End temperature

For a reliable determination, the end temperature should be approx. 30 °C above the effect to be measured. Incipient decomposition is not all that critical in this method as µTA generally does not entail defined cooling followed by a second heating.

Evaluation/ Interpretation

After the scan, the topography and thermal conductivity images are examined. The localized thermal analysis at different areas is usually evaluated in terms of temperature.

7.2.2   Influential Factors and Possible Errors During Measurement

7.2.2.1   Specimen Preparation

Before micro-TA can be performed, the sample must meet a number of conditions. Its roughness must not exceed 10 µm. This means that it generally have to be carefully prepared (microtome, polishing discs, etc.).

The sample is polished in the same way as samples to be examined under the microscope. The grinding papers and grains or polishing pastes must not extract any constituents from the sample surface during grinding and polishing. Where the study is not aimed at establishing the thermal history, it may be more appropriate to prepare a film from the melt.

Furthermore, the sample must sit plane-parallel inside the instrument. If it does not meet this specification by itself, a grinding press and modelling clay may help. Once all the conditions for measuring the sample have been met, the sample is bonded to a magnetic support and immobilized on the sample table.

A smooth, plane-parallel sample is required.

7.2.2.2   Generating the Surface Image

In preparation for the measurement, the laser beam needs to be aligned with the centers of the detectors. The tip may then be carefully positioned over the sample surface. The desired scan area is selected and then the tip is automatically placed on the sample. The surface is scanned over a maximum area of 100 x 100 µm. The following images may be recorded:

- Topography: the topography of the scanned area,
- Sensor: a continuous visualization of the internal sensor to emphasize topography features,
- Conductivity: a thermal conductivity image, generated from the power needed to keep the tip at the desired temperature,
- Mod Temp (Amp)*:   a visualization based on the response signal for a specified temperature-modulated amplitude,
- Mod Temp (Phase)*:   a visualization based on the phase shift between specified and attained temperature modulation,
- Z Piezo: the voltage [V] of the Z piezo position is visualized to emphasize topography features,

  this function is useful for visualizing very flat surfaces in which the size of a feature is less than the baseline noise of the scanner.

These visualizations only apply to modulation experiments.

The thermal conductivity and topography images of the sample are generally used to prepare for micro-TA. They aid selection of the areas for localized thermal analysis.

The thermal conductivity image is obtained by keeping the sensor at a constant temperature and measuring the change in sensor resistance. The temperature should be at least 30 °C above the sample's temperature (i.e., for room temperature: approx. 60 °C).

7.2.2.3   Choice of Measuring Points

The choice of measuring points depends on the objective. With a multilayer sample, for example, points in every layer are chosen; at an interface, they should be in each surface and in the boundary zone. As the tip usually penetrates into the sample during the measurement and leaves a small crater, the points should be at least 15 µm apart.

Between the individual measuring points, a baseline is measured in air and subtracted from the subsequent experimental curve. When melt residue clings to the tip, it must be burned off by thorough heating. A useful practical approach is to perform all the measurements in one area of the same material and to heat the tip thoroughly before measuring an area of different material or properties. This should prevent sample contamination from generating spurious results.

Sample contamination on the tip may lead to spurious results.

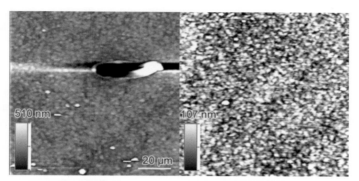

Fig. 7.5   Z-piezo image of a polished PA 6 surface before (right) and
                  after (left) a TA measurement

                  Scanning rate 25 μm/s, scanning temperature 60 °C, heating rate 20 °C/s,
                  end temperature 260 °C, load 25 nA

7.2.2.4   Load

As shown in Fig. 7.1, a mirror attached to the cantilever reflects the laser beam toward the
detector every time the tip is deflected. The mirror system and the sensor can be adjusted via
screws to make the laser impinge on the center of the detector. This ensures that large sensor
deflections, including the positive and the negative z-directions, are detected, for example,
those caused by samples with a high degree of roughness.

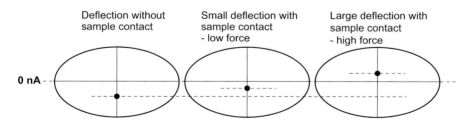

Fig. 7.6   Setting the amount of laser deflection on the photodetector before and after the tip makes
                  contact with the sample

When the tip automatically makes contact on the sample surface, the resultant deflection on
the detector can be preset. Because it is possible to adjust the reflection of the laser beam

when contact has not been made and to predetermine the position on the sample, this affords an indirect means of varying the contact force of the sensor on the sample, Fig. 7.6. However, the scope for varying the contact force is very restricted by the great flexibility of the cantilever. Therefore, it is vital to strike a compromise between ready identification of penetration into the sample during softening and ready removal again. If the tip penetrates too far, it may pull out a large amount of sample on removal and may even be damaged.

7.2.2.5   Measuring Program

Starting and end temperatures are subject to the same constraints as for classical thermal analysis. The starting temperature should be low enough to clearly detect the effect to be measured and the end temperature should be far enough above the measured effect, yet below the decomposition temperature.

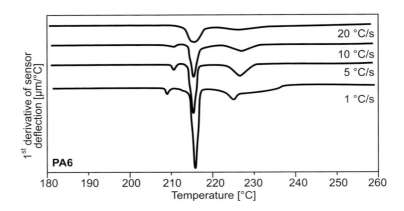

Fig. 7.7   Influence of different heating rates on the formation of melting effects in a PA 66

Load 20 nA, polished surface

The heating rates in micro-TA are substantially higher than in DSC and TMA because the sample areas are very small. However, the experimental parameters do also vary with the type of sample and the objective. A heating rate of 5–10 °C/s is suitable for determining the melting temperatures of thermoplastics. A higher rate may sometimes mask small effects (Fig. 7.7). Very low rates may visualize melting effects from beneath the surface; they are reached after a delay because heating is slow.

The shift observed in melting temperatures during their determination at high heating rates is not very pronounced, as the example of PP shows Fig. 7.8. Micro-TA often reveals broad transitions in amorphous polymers. This may have something to do with the low thermal conductivity, which is comparable to that of semicrystalline polymers, and thus somewhat delayed transfer of heat into the sample. A further cause might be the generally higher melt viscosity immediately above the glasstransition region of amorphous polymers relative to the melt viscosity immediately above the melting range of semicrystalline polymers.

Furthermore, the heating rate is seen to have a much greater effect on the glasstransition region of amorphous plastics than it does on the melting region of semicrystalline polymers. This is due to the time-dependence of the glasstransition temperature. Consequently, it is often difficult to identify amorphous materials, and it is easy to draw the wrong conclusions.

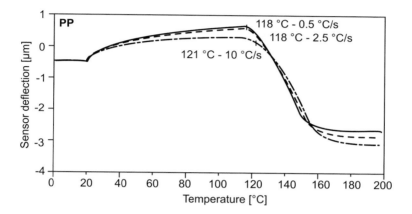

Fig. 7.8   Sensor deflection of PP at different heating rates (0.5, 2.5, 10 °C/s)

               Load 25 nA, polished surface

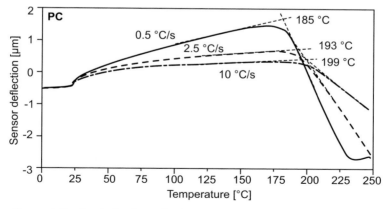

Fig. 7.9   Sensor deflection in the determination of PC at different heating rates

               Load 25 nA, polished surface

Figure 7.9 illustrates this problem with the example of PC. The glasstransition temperature of PC is usually determined to be 150 °C. However, when determined by means of µTA, it is found to be 185 °C, even at a relatively slow heating rate of 0.5 °C/s. At higher rates again, it increases to almost 200 °C. A µTA measurement should therefore never be used on its own to identify a material. Instead, it should be used to help clarify effects in materials that have already been identified.

7.2.2.6   Evaluation

Owing to the very small contact area between the sensor tip and the sample, it is essential that the experimental result be reproduced several times (at least three). Otherwise, the effect cannot be considered significant.

At least three reproductions of the experimental result are needed.

The onset temperature of the effect concerned is usually employed in the evaluation of the various curves. If, as when melting temperatures are determined, subsequent further expansion of the material is observed, the maximum of the first derivative of the sensor deflection may be used for the evaluation.

7.2.3 Real-Life Examples

7.2.3.1   Identification of Polymers

µTA is theoretically capable of identifying unknown polymers from characteristic melting and glasstransition temperatures. However, given the very small sample size and the dependence of the glasstransition temperature on the heating rate as mentioned in Section 7.2.2.5, wrong conclusions may be drawn.

Evaluation of transition temperatures is usually based on tangent intersections, an approach that relies heavily on the processing conditions and constitutes a further possible source of error.

Using the melting temperature to identify semicrystalline thermoplastics is less critical. Where the materials are unknown, however, it is always best to compare the results with those of DSC and to use micro-TA for making further selection choices.

Optical-microscope examination of the ground joint of a glass-fiber-reinforced PA 66 pipe revealed inhomogeneities. Localized thermal analysis quickly showed that the inhomogeneity was a particle not of different material, but rather of the base material PA 66, Fig. 7.10.

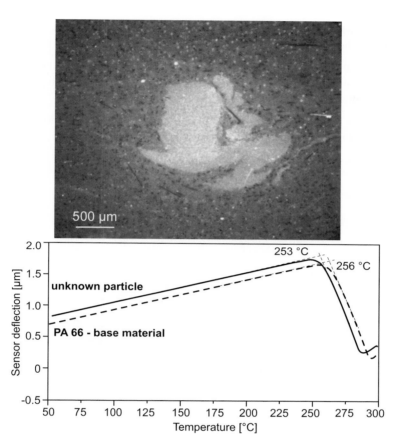

Fig. 7.10   Ground joint of a PA 66-GF pipe (top) and localized thermal analysis of base material
               and unknown particle (bottom)

Load 25 nA, heating rate 20 °C/s

7.2.3.2   Boundary Layer of a PP Sample

Under realistic processing conditions, fine spherulitic and even amorphous boundary layers
usually form on plastic moldings.

In traditional thermal analysis, the thinness of the layers makes sample preparation
laborious. A microtome is needed because the layers have to be carefully separated from
each other. Even if separation is successful, the sample is usually not large enough for
performing a DSC measurement. The fact that the µTA tip can be positioned precisely

enables the glass-transition temperatures to be accurately determined as a function of distance from the boundary.

In this example, the crosssection of a PP sample was ground and polished. The points marked in the photograph were targeted specifically and the melting temperature of the respective layer there was determined.

It can be seen that the melting temperature in the boundary layer is the lowest. The deeper layers have higher melting temperatures.

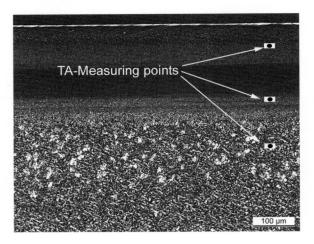

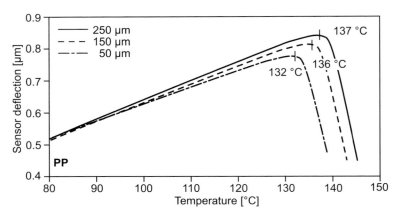

Fig. 7.11   Top: Thin section of the PP molding to illustrate the position of the
                               measuring points
               Bottom: Sensor deflection of the µTA measurement on the marked
                               positions, measured on a polished sample

               Load 25 nA, heating rate 10 °C/s

7.2.3.3 Multilayer Pipe

µTA can reveal the layer structure of a multilayer pipe (five layers) and also identify the individual materials. Each of the interfacial layers was scanned individually and measured twice on both sides. Aside from a vinyl acetate layer, which was identified by means of IR spectroscopy, the other four layers were characterized fairly reliably from their melting temperatures.

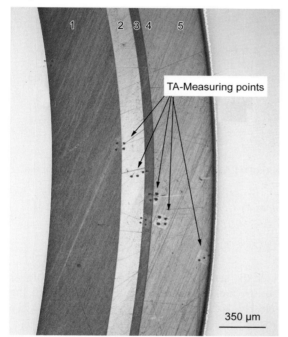

Point	Start of melting [°C]	Material
1	221–222	PA 6
2	Possible decomposition	VA*
3	216–223	PA 6
4	146–152	PP
5	184–189	PA 12

*VA was identified by means of IR

– The measuring points show up as craters

Fig. 7.12   Multilayer structure of a plastic pipe

Scanning rate 50 µm/s, sensor temperature 60 °C, load 25 nA, heating rate 10 °C/s (black dots indicate the measuring points)

When examining interfacial layers, it can be worthwhile looking at them under the microscope again after the measurements have been performed. The craters formed during the µTA allow the position of the measuring points to be checked, Fig. 7.12.

7.2.3.4 Gating Region in a Two-Component Sample

The following example shows different views of the contact area of a two-component sample (PA 6/PA 66-GF). The topography was imaged first so that a localized µTA could be conducted at specific areas. The goal was to obtain information about the material's behavior, especially in the gating edge, as a function of injection temperature and the polymers concerned. The measurements clearly show the formation of a mixed phase at the interface.

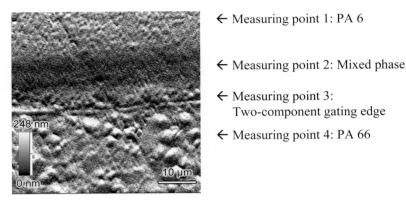

← Measuring point 1: PA 6

← Measuring point 2: Mixed phase

← Measuring point 3:
   Two-component gating edge

← Measuring point 4: PA 66

Fig. 7.13   Topography (top) and μTA (bottom) of a PA 6/PA 66 GF sample in the vicinity of the gating edge

Scanning rate 50 μm/s, sensor temperature 60 °C, load 25 nA
polished surface, heating rate 10 °C/s

A further experiment was conducted in which the sensor temperature was varied. A higher sensor temperature of 220 °C (equivalent to the melting temperature of PA 6) reveals the contact area for both materials much more clearly.

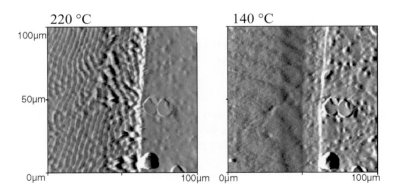

Fig. 7.14   Influence of the sensor temperature on the topography of a PA 6/PA 66 GF sample

7.2.3.5   PA 6 in a Metal Composite

Figure 7.15 shows the thermal conductivity image of a coated metal support in a composite with PA 6. The goal here was to establish the layer thickness and the miscibility of both components.

The layer thickness of the thermoset coating on the metal can be estimated from the thermal conductivity scan. Localized thermal analysis was performed at several points in the region of the polyamide and the thermoset coating.

A mixed phase consisting of both components was not detected. In the coating region, an effect at around 243 °C suggested that the thermoset coating had been postcured.

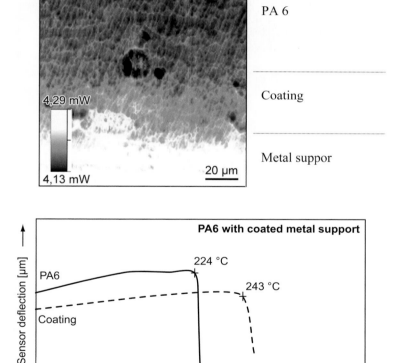

Fig. 7.15   Thermal conductivity (above) and μTA at the gating edge of a PA 6-metal composite

Scanning rate 50 μm/s, sensor temperature 60 °C, load 25 nA
polished surface, heating rate 10 °C/s

7.2.3.6   Detection of Surface Aging

The following two 3D topography images show the boundary layers of injection-molded PA 66 panels that had been conditioned at various temperatures for different periods of time. A boundary layer that is low in spherolite is clearly visible in both cases. Aging at high temperatures (120 °C) leads to an increase in crystallinity and is clearly linked to an enlargement of spherolitic structures. These effects show up in the localized thermal analysis as a rise of several degrees Celsius in melting temperature of the postcured sample, both in the boundary layer and in the sample bulk.

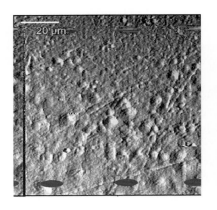 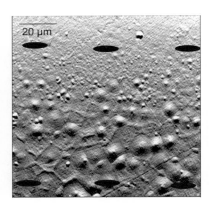

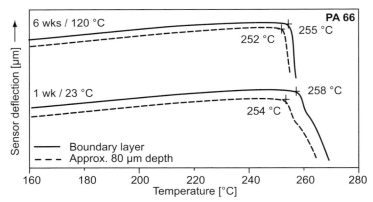

Fig. 7.16   Top: 3D topography images
               Bottom: µTA measurements of injection-molded PA 66 panels after
                               different aging periods
                               (above left: 1 week/23 °C; above right: 6 weeks/120 °C)

               Scanning rate 50 µm/s, sensor temperature 60 °C, load 25 nA
               polished surface, heating rate 10 °C/s

7.3 References

[1] zur Mühlen, E.

Introduction to Microthermal Analysis and its
Application to the Study of Elastomers
Proceedings of Elastomer Analysis, Deutsches Institut
für Kautschuktechnologie e.V., Hannover 1998

[2] N.N.

TA-Instruments, µTA 2290 Micro Thermal Analyzer
Operator's Guide
July 2000

Other References:

[3] Häßler, R.,
zur Mühlen, E.

An Introduction to µTA™ and Application to the
Study of Interfaces
Thermochimica Acta 361(2000), pp. 113–120

[4] Häßler, R.

Mikrothermische Analyse an Grenzschichten
kleben & dichten Adhäsion 43 (1999) 6, pp. 30–33

[5] Price, D. M.

http://www.sump4.com/publications

[6] Moon, I.,
Androsch, R.,
Chen, W.,
Wunderlich, B.

The Principles of Micro-thermal Analysis and its
Application to the Study of Macromolecules
Journal of Thermal Analysis and Calorimetry
59 (2000) 1, pp. 187–203

[7] Price, D. M.,
Reading, M.,
Hammiche, A.,
Pollock, H. M.

New Adventures in Thermal Analysis
Journal of Thermal Analysis and Calorimetry
60 (2000) 3, pp. 723–733

[8] Price, D. M.,
Lever, T. J.
Reading, M.,

Applications of Micro-thermal Analysis
Journal of Thermal Analysis and Calorimetry
56 (1999) 2, pp. 673–679

[9] Tsukruk, V.,
Gorbunov, V.,
Fuchigami, N.

Microthermal Analysis of Polymeric Materials
Thermochimica Acta 395 (2003), pp. 151–158

8   Brief Characterization of Key Polymers

There are many structural ways in which the properties of polymers can be modified, such as via the molecular weight, crystallization, copolymerization, reinforcement, stabilization, and so forth. Processing also has a major impact on them – more so, in fact, than it does on conventional materials. The characteristics listed on the following pages broadly apply to standard materials, with the selection biased toward properties needed in applications. Listed are the **microstructure**, the most important **service properties**, the **chemical structure**, and a selection of **physical characteristics**. The glass transition temperature, T_g, was determined by means of DSC (heating rate: 10 °C/min) and evaluated as the midpoint temperature, T_{mg}. T_{pm} denotes the peak temperature of the melting range measured in DSC. Values for **service limits** are based on literature references and on our own experience. There is no suitable way of measuring these directly. The mechanical properties were determined in the static tensile test as set out in ISO 527 and comparable norms. The **prices** shown are intended only as a rough guide; they are the average values for medium-size quantities of class 1A goods in 2002.

The $\sigma - \varepsilon$ **diagram** characterizes the strength, deformation, stiffness, and energy-absorption behavior as determined in the tensile test. With the exception of continuous fiber-reinforced materials, such "tensile values" are generally lower than their compressive or flexural stress counterparts.

The curve of the **elastic modulus** against temperature was obtained by means of DMA under very slight torsional or flexural loads. Deviations from the "static tensile" values arise from differences in sample geometry, type of loading, and the level of stress.

The **DSC diagrams** (second heating) of amorphous materials were obtained at a heating rate of 20 °C/min for a specimen mass of approx. 10 mg, with the corresponding figures for semicrystalline materials being 10 °C/min and approx. 3 mg. In each case, the measuring cell was purged with nitrogen and the sample cooled prior to measurement at a defined cooling rate (which was the same as the heating rate) from above the transition temperature.

The curve of the **coefficient of expansion** against temperature was determined by means of TMA. The heating rate was 3 °C/min, the applied load was 5 g, and the piston had a diameter of 2 mm. As with the DSC measurements, the thermal history of the TMA samples was eliminated by prior heating and subsequent defined cooling, with the cooling rate matching the subsequent heating rate. All samples were measured parallel to the direction of injection.

ABS – Acrylonitrile-Butadiene-Styrene Copolymer

Type: amorphous, blend   **Price (2002):** ≈ 1.70 €/kg

Characteristics: shiny surface, stiffness, scratch resistance, high dimensional stability, impactresistance, ability to be galvanized, poor environmental resistance

Identified by: opaqueness, smells sweet (styrene) when burned, bright flame, soot forming

Properties:
ρ = 1.03–1.07 g/cm³
E = 2200–3000 MPa
σ_s = 45–65 MPa
ε_s = 2.5–3%
ε_B = 15–20%
T_g = -85 °C/ 95–105 °C

Thermal limits:
Short time: ~ 85–95 °C
Long time: ~ 75–85 °C

Structure:

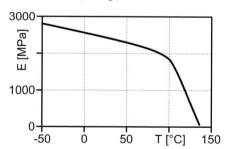

Matrix: $-\!\!\left[CH_2\!-\!CH\right]_n\!\!\left[CH_2\!-\!CH\right]_m\!\!-$ (with phenyl ring on first unit, CN on second)

Rubber: $-\!\!\left[CH_2\!-\!CH\!=\!CH\!-\!CH_2\right]_o\!\!-$

Often blended with:
PA, TPU, PVC, PC, PSU, PUR

σ - ε - Diagram:

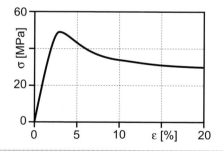

E-Modulus f (Temp):

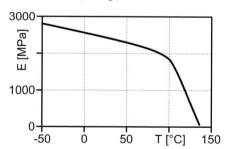

DSC - Diagram:

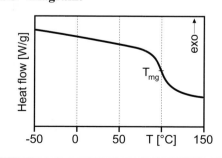

Coeff. of linear therm. expansion f (Temp):

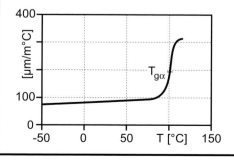

ASA – Acrylonitrile-Styrene-Acrylester Copolymer

Type: amorphous   **Price (2002):** ≈ 2.40 €/kg

Characteristics: toughness, stiffness, shiny surface, higher enviromental and chemical resistance than ABS

Identified by: opaqueness, smells sweet (styrene) when burned, bright flame, soot forming

Properties:
ρ = 1.04–1.07 g/cm^3
E = 2300–2900 MPa
σ_s = 40–55 MPa
ε_s = 3.1–4.3%
ε_B = 10–30%
T_g = -40 °C / 95 °C

Thermal limits:
Short-time: ~ 85–95 °C
Long-time: ~ 75–85 °C

Structure:

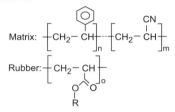

Often blended with:
PC, PVC, PMMA

σ - ε - Diagram:

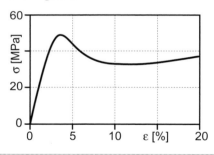

E-Modulus f (Temp):

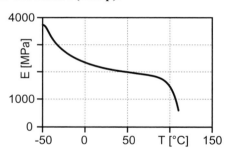

DSC - Diagram:

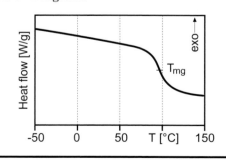

Coeff. of linear therm. expansion f (Temp):

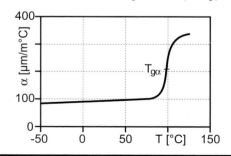

COC – Cycloolefin Copolymer

Type:	amorphous **Price (2002):** ≈ 3–10 €/kg
	Copolymer - linear polyolefinenes and norbornems
	(cycl. polyolefines)
Characteristics:	high transparency, low water absorption, predominantly stiff and brittle, resistance against acids and bases, low density
Identified by:	fiber spun melt, smell of pungent paraffin when burned, smoke has pH-value of 4

Properties*:

ρ = 1.02 g/cm^3
E = 2600–3200 MPa
σ_B = 66 MPa
ε_B = 3–10%
T_g = 0–230 °C

Thermal limits:

Short-time: ~ 75–170 °C
Long-time: ~ 65–100 °C

* dependent on cyclo-olefin concentration

Structure:

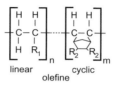

linear
olefine
cyclic

σ - ε - Diagram:

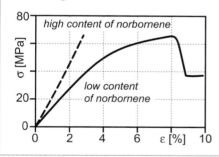

E-Modulus f (Temp):

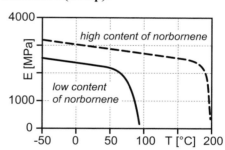

DSC - Diagram:

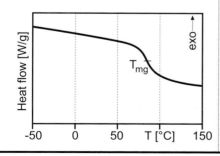

Coeff. of linear therm. expansion f (Temp):

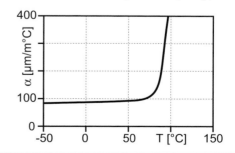

EP – Epoxy Resin

Type: amorphous, crosslinked   **Price (2002):** ≈ 5.00–8.00 €/kg

Characteristics: large variability, often reinforced, toughness to brittleness, potential for many modifications (e.g., rubber), high dimensional stability, durable, good electrical insulation

Identified by: transparency to opaqueness, does not melt

Properties*:
ρ = 1.17–1.25 g/cm^3
E = to 4200 MPa
σ_B = to 100 MPa
ε_B = ~ 1.5–20%
T_g = ca. 70–200 °C

Thermal limits:
Different for each type
Long-time up to 60/180 °C
(depends on T_g)
*for nonreinforced material

Structure:

$$R_1-NH_2^+ \ CH_2-CH-R_2 \rightarrow R_1-N-CH_2-CH-R_2$$

hardener + epoxy group → EP-resin

Often blended with:
polysulfid-rubber, tar products

σ - ε - Diagram:

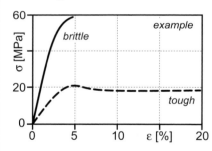

E-Modulus f (Temp):

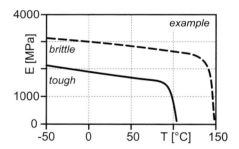

DSC - Diagram:

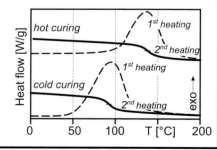

Coeff. of linear therm. expansion f (Temp):

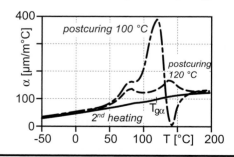

PA 11 – Polyamide 11

Type:	semicrystalline (35–45%)	**Price (2002):** ≈ 7.50 €/kg
Characteristics:	high toughness, low water absorption (standard humidity 1.1%, saturated 1.8%)	
Identified by:	milky white-yellow color, smell of burnt horn when burned, yellow flame with a blue halo, can be formed into a filament, melt drips	

Properties:
ρ = 1.03–1.05 g/cm^3
E = 1400/1200/– MPa
σ_s = 45/40/– MPa
ε_s = 5–10/10 - 30/- %
ε_B = > 50%
T_g = 49/–/– °C
T_{pm}= 185 °C

Thermal limits:
Short-time: ~ 140 °C
Long-time: ~ 70–80 °C

Structure:

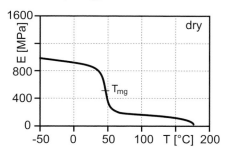

Often blended with:
PA 6, PA 66

σ - ε - Diagram:

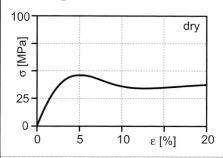

E-Modulus f (Temp):

DSC - Diagram:

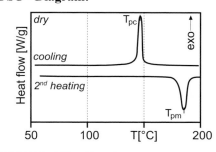

Coeff. of linear therm. expansion f (Temp):

—

PA 12 – Polyamide 12

Type:	semicrystalline (35–45%)	**Price (2002):** ≈ 7.50 €/kg
Properties:	low density, resistance to stress cracking, low water absorption (standard 0.7%, saturated 1.5%)	
Identified by:	milky white-yellow color, smell of burnt horn when burned, yellow flame with a blue halo, can be formed into a filament, melt drips	

Properties:
ρ = 1.01–1.04 g/cm^3
E = 1400/1100/570 MPa
σ_s = 50/40/- MPa
ε_s = 4/12/- %
ε_B = ~ 200%
T_g = 49/-/- °C
T_{pm}= 170–180 °C

Thermal limits:
Short-time: ~ 140 °C
Long-time: ~ 70 - 80 °C

Structure:

$$\left[NH-(CH_2)_{11}-\overset{\displaystyle O}{\overset{\displaystyle \|}{C}} \right]_n$$

Often blended with:
PA 6, PA 66

σ - ε - Diagram:

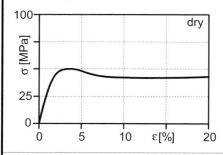

E-Modulus f (Temp):

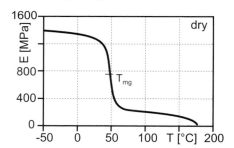

DSC - Diagram:

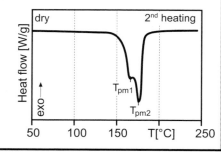

Coeff. of linear therm. expansion f (Temp):

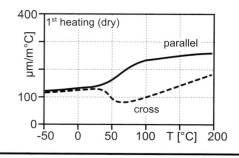

PA 46 – Polyamide 46

Type: semicrystalline (60–70%)   **Price (2002):** ≈ 6.00 €/kg

Characteristics: abrasion resistance, high thermal stability, low creep, absorption of water up to 12%

Identified by: milky white-yellow color, smell of burnt horn when burned, yellow flame with a blue halo, can be formed into a filament

Properties:

ρ = 1.18 –1.21 g/cm^3
E = 3300/1000/800 MPa
σ_s = 100/55/– MPa
ε_s = –/–/–
ε_B = 40/> 200/- %
T_g = ~ 94/31/-10 °C
T_{pm}= 285–290 °C

Thermal limits:

Short-time: ~ 180 °C
Long-time: ~ 120 °C

Structure:

$$\left[NH-(CH_2)_4-NH-\overset{O}{\overset{\|}{C}}-(CH_2)_4-\overset{O}{\overset{\|}{C}} \right]_n$$

Often blended with:

PTFE

σ - ε - Diagram:

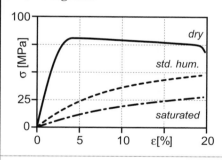

E-Modulus f (Temp):

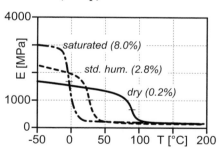

DSC - Diagram:

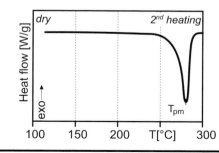

Coeff. of linear therm. expansion f (Temp):

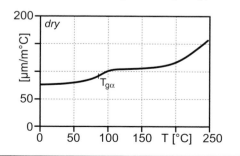

PA 6 – Polyamide 6

Type: semicrystalline (30–40%)   **Price (2002):** ≈ 2.50 €/kg

Charakteristics: abrasion resistance, toughness, multifunctional, absorption of water up to 9.5%, often reinforced

Identified by: milky white-yellow color, smell of burnt horn when burned, yellow flame with a blue halo, can be formed into a filament

Properties:
ρ = 1.12–1.15 g/cm^3
E = 2800/1000/600 MPa
σ_s = 80/45/– MPa
ε_s = 4/25/– %
ε_B = 30/> 50/– %
T_g = ~ 78/28/-8 °C
T_{pm}= 225–235 °C

Thermal limits:
Short-time: ~ 140–160 °C
Long-time: ~ 80–100 °C

Structure:

$$\left[NH-(CH_2)_5-\overset{\displaystyle O}{\overset{\displaystyle \|}{C}} \right]_n$$

Often blended with:
PA 66, PA 12, PA 11, PE, ABS, ABR, PPS, PPE, PTFE, EPDM

σ - ε - Diagram:

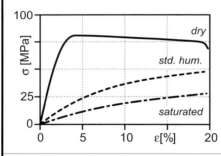

E-Modulus f (Temp):

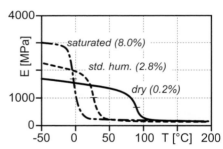

DSC - Diagram:

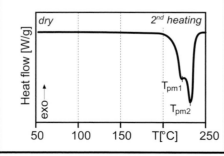

Coeff. of linear therm. expansion f (Temp):

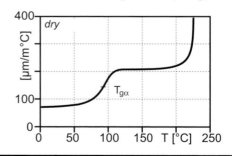

PA 610 – Polyamide 610

Type: semicrystalline (30–40%)   **Price (2002):** ≈ 6.00 €/kg

Characteristics: low water absorption (3.3%), good dimensional stability

Identified by: milky white-yellow color, smell of burnt horn when burned, yellow flame with a blue halo, can be formed into a filament

Properties:

ρ = 1.07–1.09 g/cm^3
E = 2200/1500/800 MPa
σ_s = 64/45/- MPa
ε_s = 4/15/- %
ε_B = > 50%
T_g = 77/48/19 °C
T_{pm} = 210–220 °C

Thermal limits:
Short-time: ~ 140–160 °C
Long-time: ~ 80–100 °C

Structure:

$$\left[NH-(CH_2)_6-NH-\overset{O}{\overset{\|}{C}}-(CH_2)_8-\overset{O}{\overset{\|}{C}} \right]_n$$

Often blended with:
PA 66/6

σ - ε - Diagram:

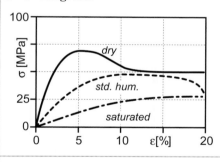

E-Modulus f (Temp):

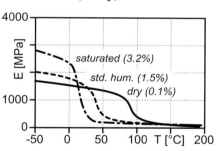

DSC - Diagram:

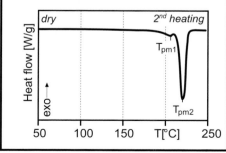

Coeff. of linear therm. expansion f (Temp):

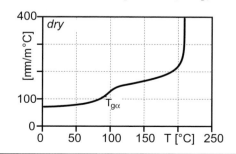

PA 6-3-T – Amorphous Polyamide

Type: amorphous   **Price (2002):** ≈ 8.00 €/kg

Characteristics: toughness, absorption of water up to 7.5%

Identified by: transparency, smell of burnt horn when burned, yellow flame
                         with a blue halo, can be formed into a filament, melt drips

Properties:

ρ = 1.12–1.15 g/cm^3
E = 2800/2700/2200 MPa
σ_s = 90/85/65 MPa
ε_s = 7/6/5 %
ε_B = 20 – > 50%
T_g = 152/114/97 °C

Thermal limits:
Short-time: ~ 120 °C
Long-time: ~ 80 °C

Structure:

$$\left[NH-(CH_2)_6-NH-\overset{O}{\overset{\|}{C}}-\langle\bigcirc\rangle-\overset{O}{\overset{\|}{C}} \right]_n$$

σ - ε - Diagram:

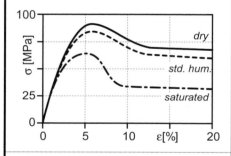

E-Modulus f (Temp):

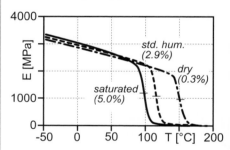

DSC – Diagram:

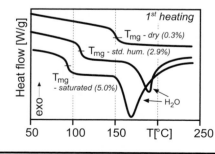

Coeff. of linear therm. expansion f (Temp):

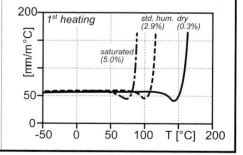

PA 66 – Polyamide 66

Type: semicrystalline (35–45%)   **Price (2002):** ≈ 3.00 €/kg

Characteristics: high abrasion resistance, tough, aging and thermal resistance, absorption of water up to 8.5%, often reinforced

Identified by: transparency, smell of burnt horn when burned, yellow flame with a blue halo, can be formed into a filament

Properties:	**Structure:**

Properties:

ρ = 1.13–1.16 g/cm^3

E = 3000/1600/800 MPa

σ_s = 85/60/– MPa

ε_s = 5/20/– %

ε_B = 25/> 50/– %

T_g = ~ 90/39/-6 °C

T_{pm}= 225–265 °C

Thermal limits:

Short-time: ~ 140–170 °C

Long-time: ~ 80–100 °C

Structure:

$$\left[NH-(CH_2)_6-NH-\overset{O}{\overset{\|}{C}}-(CH_2)_4-\overset{O}{\overset{\|}{C}} \right]_n$$

Often blended with:

PA 6, PA 12, PA 11, PE, ABS, ABR, PPS, PPE, PTFE, EPDM

σ - ε - Diagram:

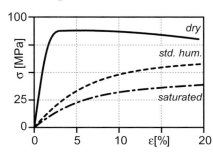

E-Modulus f (Temp):

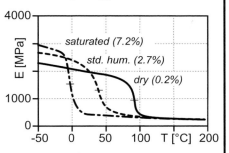

DSC - Diagram:

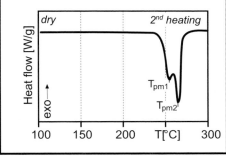

Coeff. of linear therm. expansion f (Temp):

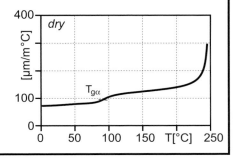

PAI – Polyamidimid

Type: amorphous, crosslinked   **Price (2002):** ≈ 6.00 €/kg

Characteristics: stiff, hard, low toughness, good skip and war properties, low thermal
                      expansion, mediumenvironmental resistance, 3.5%–4% water uptake

Identified by: –

Properties:
ρ = 1.38–1.40 g/cm^3
E = 4500–5200 MPa
σ_s = –
ε_s = –
ε_B = 10%
T_g = 250 °C–275 °C

Thermal Limits:
Short-time: ~ 300 °C (type-dependent)
Long-time: ~ 260 °C (type-dependent)

Structure:

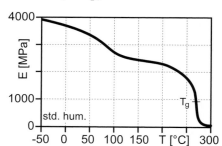

Often blended with:
PTFE, PSU, PEI, PA, PPS, PC

σ - ε - Diagram:

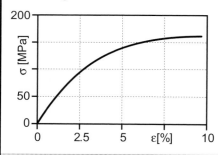

E-Modulus f (Temp):

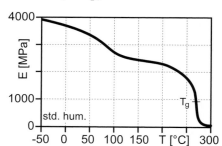

DSC - Diagram:

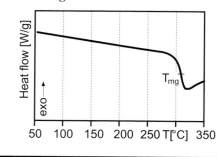

Coeff. of linear therm. expansion f (Temp):

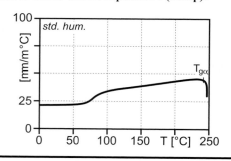

PB - Polybutene

Type: semicrystalline (ca. 50%)   **Price (2002):** ≈ 3.00 €/kg

Characteristics: soft and elastic after injection moulding, later ($\sim$ 7 days) stiffer and ductil, low density good stresscracking resistance, low creep

Identified by: burns with yellow flame, blistering, dipping flame

Properties:

ρ = 0.89* / 0.91– 0.94 g/cm^3

E = 240* / 600–700 MPa

σ_s = 15–25 MPa

ε_s = 10%

ε_B = > 50%

T_{pm}= 130 °C

Thermal Limits:

Short-time: $\sim$ 80–110 °C

Long-time: $\sim$ 60–60 °C

* Data immediately after moulding

Structure:

$$\left[CH_2 - \underset{\underset{CH_3}{\overset{|}{CH_2}}}{\overset{|}{CH}} \right]$$

Often blended with:

- PE, - PP, - PS

σ - ε - Diagram:

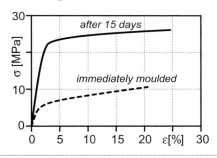

E-Modulus f (Temp):

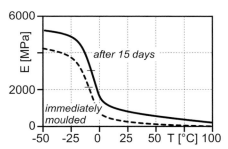

DSC - Diagram:

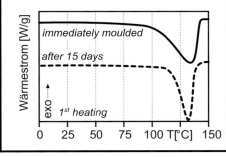

Coeff. of linear therm. expansion f (Temp):

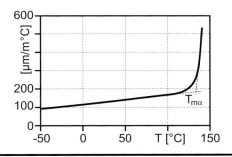

PBT – Polybutylene Terephthalate

Type: semicrystalline (40 - 50%)   **Price (2002):** ≈ 2.80 €/kg

Characteristics: good dimensional stability, sensitivity to processing, favorable sliding and wear characteristics, good insulating properties, often reinforced

Identified by: orange colored sooty flame, smells sweet when burned, melt drips

Properties:
ρ = 1.30–1.32 g/cm^3
E = 2500–2800 MPa
σ_s = 50–60 MPa
ε_s = 3,5–7%
ε_B = 20–> 50%
T_g = 45–60 °C
T_{pm}= 220–230 °C
Thermal limits:
Short-time: ~ 160 °C
Long-time: ~ 100 °C

Structure:

$$\left[O-(CH_2)_4-O-\overset{O}{\underset{\|}{C}}-\bigcirc-\overset{O}{\underset{\|}{C}} \right]_n$$

Often blended with:
 ASA, EPDM, LCP, PC, PET

σ - ε - Diagram:

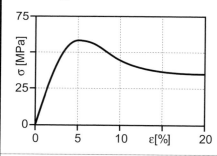

E-Modulus f (Temp):

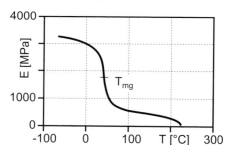

DSC - Diagram:

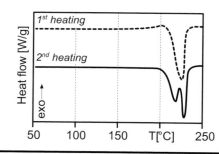

Coeff. of linear therm. expansion f (Temp):

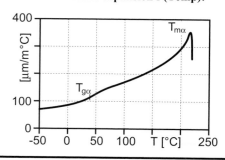

PC – Polycarbonate

Type:	amorphous **Price (2002):** ≈ 3.40 €/kg
Characteristics:	good dimensional stability, shiny surface, impactresistance, high thermal stability, sensitivity to stress cracking, good insulating properties
Identified by:	transparency, dark yellow flame, soot forming, self extinguishing

Properties:

ρ = 1.20–1.24 g/cm^3
E = 2200–2400 MPa
σ_s = 55–65 MPa
ε_s = 6–7%
ε_B = 100–130%
T_g = 145 °C

Thermal limits:

Short-time: ~ 135 °C
Long-time: ~ 100 °C

Structure:

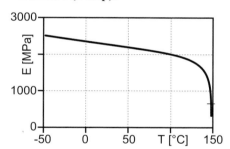

Often blended with:

ABS, ASA, SEBS, PBT, PET, PPE, SB, PS, PPE, PP, TPU

σ - ε - Diagram:

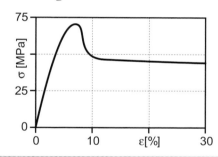

E-Modulus f (Temp):

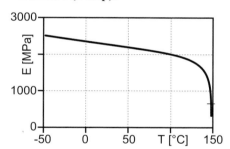

DSC – Diagram:

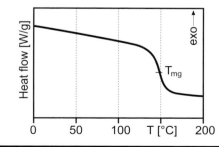

Coeff. of linear therm. expansion f (Temp):

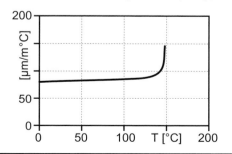

PC-ABS – PC-ABS-Blend

Type:	amorphous, blend **Price (2002):** ≈ 3.00 €/kg
Characteristics:	high dimensional stability, good thermal stability, impactresistance, ability to galvanized
Identified by:	opaqueness, smells sweet (styrene) when burned, bright flame, soot forming

Properties:
ρ = 1.08–1.17 g/cm^3
E = 2000–2600 MPa
σ_s = 40–60 MPa
ε_s = 3.0–3.5%
ε_B = > 50%
T_g = -85/105/145 °C
Thermal limits:
Short-time: ~ 115–130 °C
Long-time: ~ 105–115 °C

Structure:

———

σ - ε - Diagram:

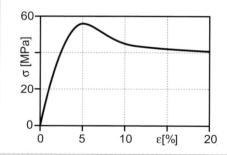

E-Modulus f (Temp):

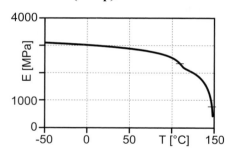

DSC – Diagram:

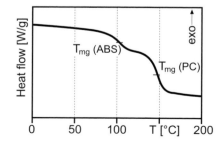

Coeff. of linear therm. expansion f (Temp):

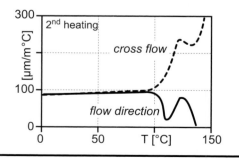

PEEK – Polyetherether Ketone

Type: amorphous or semicrystalline (~ 35%)   **Price (2002):** ≈ 80.00 €/kg

Characteristics: strength, stiffness, good stress cracking resistance (except with acetone), high thermal and chemical resistance, compatible with PEI

Identified by: opaqueness, difficult to ignite, low smoke development, slight smell of phenol when burned

Properties:

ρ = 1.32 g/cm^3 (cry.), 1.27 g/cm^3 (am.)
E = 3700 MPa
σ_s = 100 MPa
ε_s = 5%
ε_B = > 50%
T_g = 145 °C
T_{pm} = 335 °C

Thermal limits:

Short-time: ~ 300 °C
Long-time: ~ 250 °C

Structure:

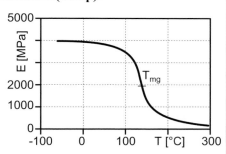

Often blended with:

PEI

σ - ε - Diagram:

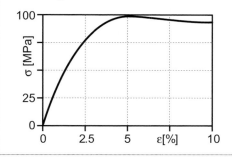

E-Modulus f (Temp):

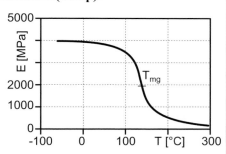

DSC - Diagram:

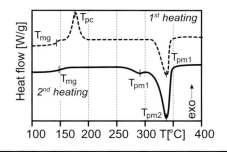

Coeff. of linear therm. expansion f (Temp):

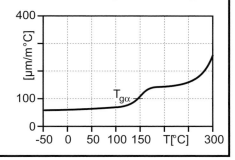

PE-HD – Polyethylene, High-Density

Type:	semi-crystalline (60– 80%)	**Price (2002):** ≈ 0.80 €/kg

Characteristics: low density, low strength and stiffness, good chemical resistance, susceptibility to photooxidation and therefore often stabilized, ability to be welded, inexpensive

Identified by: milky white color, smell of paraffin when burned, yellow flame, melt drips, floats in water

Properties:
ρ = 0.94–0.96 g/cm^3
E = 600–1400 MPa
σ_s = 18–30 MPa
ε_s = 8–12%
ε_B = > 50%
T_g = < -100 °C
T_{pm} = 125–135 °C
Thermal limits:
Short-time: ~ 80–110 °C
Long-time: ~ 60– 80 °C

Structure:

Often blended with:
PP, PA

σ - ε - Diagram:

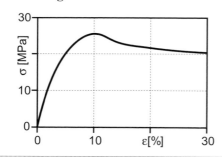

E-Modulus f (Temp):

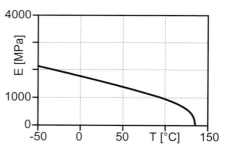

DSC - Diagram:

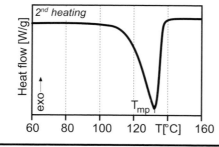

Coeff. of linear therm. expansion f (Temp):

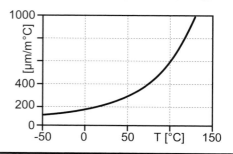

PEI – Polyetherimide

Type:	amorphous	**Price (2002):** ≈ 14.00 €/kg

Characteristics: high thermal stability, high strength, high environmental stability, low thermal expansion coefficient, compatible with PEEK, impact resistance

Identified by: transparency, amber-yellow color, smell of burnt horn when burned, yellow flame

Properties:

ρ = 1.27 g/cm^3
E = 2900–3000 MPa
σ_s = 85 MPa
ε_s = 6–7%
ε_B = > 50%
T_g = 215–230 °C

Thermal limits:
Short-time: ~ 190 °C
Long-time: ~ 160 °C

Structure:

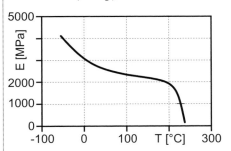

Often blended with: PEEK, PAEK

σ - ε - Diagram:

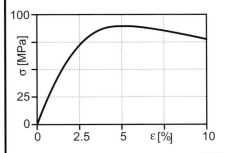

E-Modulus (Temp):

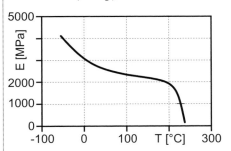

DSC - Diagram:

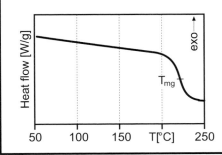

Coeff. of linear therm. expansion f (Temp):

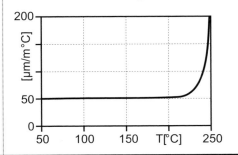

PE-LD – Polyethylene, Low-Density

Type:	semicrystalline (40–55%)	**Price (2002):** ≈ 0.90 €/kg

Characteristics: low density, easily processed by several methods, inexpensive, good flow properties, low dimensional stability

Identified by: translucence, smell of paraffin when burned, yellow flame, melt drips, floats in water

Properties:
ρ = 0.914–0.928 g/cm^3
E = 200–400 MPa
σ_s = 8–10 MPa
ε_s = ~ 20%
ε_B = > 50%
T_g = < -100/ -10 °C
T_{pm} = 100–110

Thermal limits:
Short-time: ~ 80–90 °C
Long-time: ~ 60–70 °C

Structure:

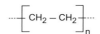

Often blended with:
PE-HD, PP, PA

σ - ε - Diagram:

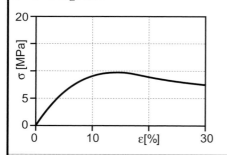

E-Modulus f (Temp):

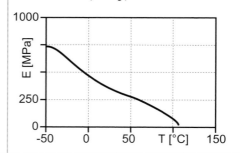

DSC – Diagram:

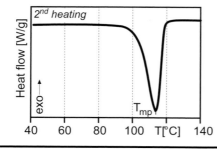

Coeff. of linear therm. expansion f (Temp):

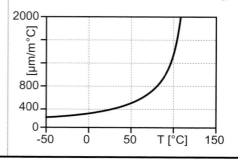

PES – Polyethersulfone

Type:	amorphous

Price (2002): ≈ 15.00 €/kg

Characteristics: high thermal stability, good chemical resistance,
susceptibility to crack formation, water absorption

Identified by: transparency, light yellow color, difficult to ignite

Properties:
ρ = 1.37 g/cm^3
E = 2600–2800 MPa
σ_s = 80–90 MPa
ε_s = 5.5–6.5%
ε_B = 20–80%
T_g = 225–230 °C
Thermal limits:
Short-time: ~ 210 °C
Long-time: ~ 180 °C

Structure:

σ - ε - Diagram:

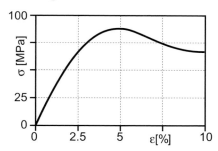

E-Modulus f (Temp):

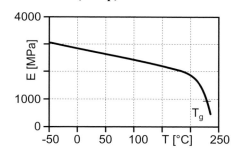

DSC – Diagram:

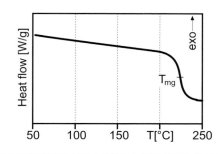

Coeff. of linear therm. expansion f (Temp):

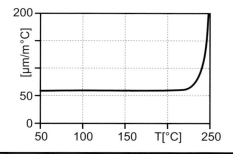

PET –Polyethylene Terephthalate

Type: amorphous, semicrystalline (30–40%)   **Price (2002):** ≈ 2.00 €/kg

Characteristics: normally amorphous and high transparency, crystalline PET with good dimensional stability, often reinforced, CO_2-impermeable, scratch resistance, higher environmental and thermal resistance as well as more brittle than PBT

Identified by: smells sweet when burned, orange flame, soot forming, melt drips

Properties*:
ρ = 1.38–1.40 g/cm^3
E = 2100–3100 MPa
σ_s = 55–80 MPa
ε_s = 4–7%
ε_B = > 50%
T_g = 70–80 °C
T_{pm}= 250–260 °C
Thermal limits:
Short-time: ~ 170 °C
Long-time: ~ 100 °C   * data forPET-C

Structure:

$$\cdots\left[O-CH_2-CH_2-O-\overset{\overset{\displaystyle O}{\|}}{C}\left\langle\bigcirc\right\rangle\overset{\overset{\displaystyle O}{\|}}{C}\right]_n\cdots$$

Often blended with:
ASA, EPDM, LCP, PC, PBT

σ - ε - Diagram:

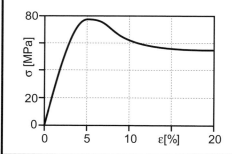

E-Modulus f (Temp):

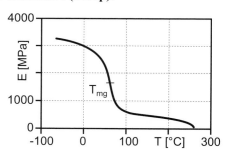

DSC – Diagram:

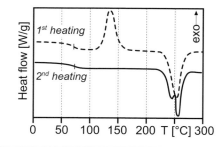

Coeff. of linear therm. expansion f (Temp):

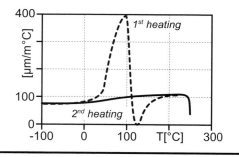

PK – Aliphatic Polyketone

Type:	semicrystalline (30–40%)	**Price (2002):** ≈ 10.00 €/kg

Characteristics: elastic, excellent impact behavior even at low temperatures, good wear characteristics, chemical and hydrolytic stability, difficult to ignite

Identified by: opaqueness, burns bright, yellow flame, forms soot, forms bubbles when melted, smoke has a pH value of 4

Properties:
ρ = 1.24 g/cm^3
E = 1400–2100 MPa
σ_s = 60–80 MPa
ε_s = 25–40%
ε_B = > 50%
T_g = 19 °C
T_{pm} = 220 °C
Thermal limits:
Short-time: ~ 180 °C
Long-time: ~ 120–140 °C

Structure:

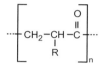

σ - ε - Diagram:

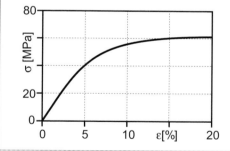

E-Modulus f (Temp):

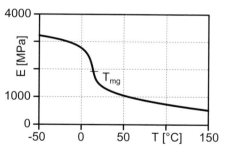

DSC – Diagram:

—

Coeff. of linear therm. expansion f (Temp):

—

PMMA – Polymethylmethacrylate

Type: amorphous   **Price (2002):** ≈ 2.60 €/kg

Characteristics: scratch resistance, stiffness, brittleness, good chemical and
                       environmental resistance, susceptible to stress cracking

Identified by: transparency, flammability, fruity and sweet smell when burned

Properties:

ρ = 1.15–1.19 g/cm^3

E = 3100–3300 MPa

σ_B = 60–80 MPa

ε_s = –

ε_B = 2–6%

T_g = 105–120 °C

Thermal limits:

Short-time: ~ 85–95 °C

Long-time: ~ 65–80 °C

Structure:

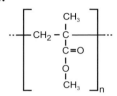

Often blended with:

ABS

σ - ε - Diagram:

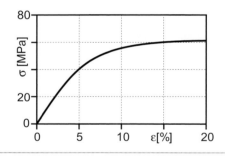

E-Modulus f (Temp):

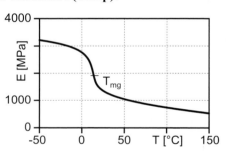

DSC – Diagram:

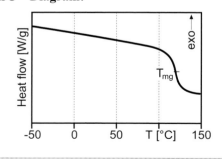

Coeff. of linear therm. expansion f (Temp):

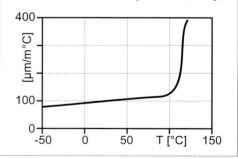

POM – Polyoxymethylene (Polyacetal)

Type:	semicrystalline **Price (2002):** ≈ 2.50 €/kg
	(homopolymer 70–80%; copolymer 50– 60%)
Characteristics:	high strength and stiffness, high dimensional stability, low susceptibility to stress cracking, sensitive to UV light
Identified by:	opaqueness, white color, strong smell of formaldehyde when burned, faint blue flame, melt drips

Properties:
ρ = 1.39–1.43 g/cm^3 (H>C)
E = 2600–3200 MPa (H>C)
σ_s = 60–75 MPa
ε_s = 8–25%
ε_B = 20–> 50% (H); 15–40% (C)
T_g = -70 °C
T_{pm}= 175 °C (H); 164–172 °C (C)

Thermal limits:

Short-time: ~ 110–140 °C
Long-time: ~ 90–100 °C

Stucture:

homopolymer

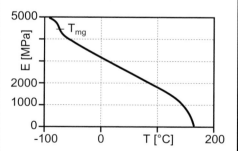

copolymer

Often blended with:
TPU, PTFE

σ - ε - Diagram:

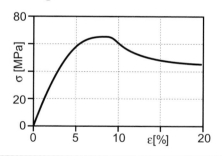

E-Modulus f (Temp):

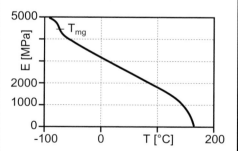

DSC – Diagram:

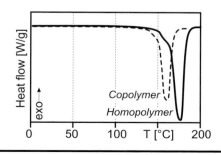

Coeff. of linear therm. expansion f (Temp):

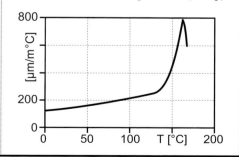

PP-H – Polypropylene (Homopolymer, isotactic)

Type:	semicrystalline (60–70%)	**Price (2002):** ≈ 0.90 €/kg

Characteristics: low cost, low stiffness, low strength, low dimensional stability, low stress cracking even at low temperatures, requires stabilization, multifunctional, often filled and reinforced, chemical resistance

Identified by: translucent, smell of paraffin when burned, bright flame, melt drips, floats in water

Properties:
ρ = 0.90–0.91 g/cm^3
E = 1300–1800 MPa
σ_s = 25–40 MPa
ε_s = 8–18%
ε_B = > 50%
T_g = 0–20 °C
T_{pm}= 160–165 °C
Thermal limits:
Short-time: ~ 130 °C
Long-time: ~ 90 °C

Structure:

$$\cdots\left[CH_2-CH\atop\quad\ |\atop\quad CH_3\right]_n\cdots$$

Often blended with:
PE, EPDM

σ - ε - Diagram:

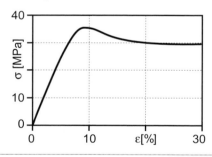

E-Modulus f (Temp):

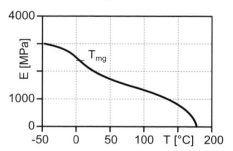

DSC - Diagram:

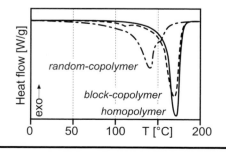

Coeff. of linear therm. expansion f (Temp):

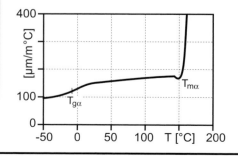

PPO+PS-HI – Polyphenylene Oxide + Impact-Resistant Polystyrene

Type: amorphous, blend   **Price (2002):** $\approx$ 3.50 €/kg

Characteristics: hardness, stiffness, impact resistance, scratch resistance, good dimensional stability, good stress crack resistance

Identified by: opaqueness, self extinguishing, melt does not drip

Properties:

ρ = 1.04–1.06 g/cm^3

E = 2300 MPa

σ_s = 50–55 MPa

ε_s = 3–5%

ε_B = 36–45%

T_g = 140 °C

Thermal limits:

Short-time: ~ 120 °C

Long-time: ~ 100 °C

Structure:

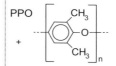

PPO + ... (structure)

PS-HI ...

σ - ε - Diagram:

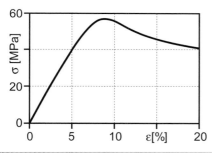

E-Modulus f (Temp):

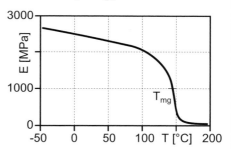

DSC - Diagram:

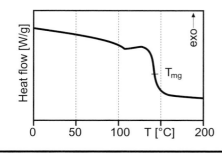

Coeff. of linear therm. expansion f (Temp):

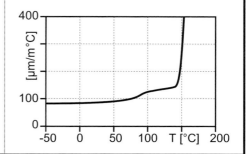

PPS – Polyphenylene Sulfide

Type:	semicrystalline	**Price (2002):** ≈ 8.00 €/kg

Characteristics: high brittleness, stiffness, hardness, high thermal, chemical, and environmental resistance, often reinforced

Identified by: opaqueness, difficult to ignite, smell of hydrogen sulfide when burned

Properties*:
ρ = 1.34–1.36 g/cm^3
E = 3700 MPa
σ_s = 75 MPa
ε_B = 4%
T_g = 85–95 °C
T_{pm} = 285–290 °C

Thermal limits:
Short-time: 270–300 °C
Long-time: 200–240 °C

*for nonreinforced material

Stucture:

σ - ε - Diagram:

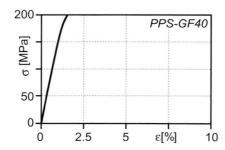

E-Modulus f (Temp):

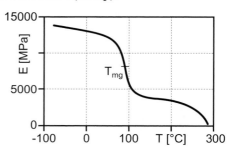

DSC – Diagram:

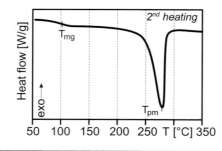

Coeff. Of linear therm. Expansion f (Temp):

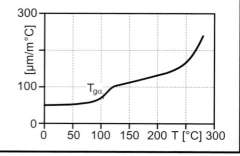

PS – Polystyrene

Type:	amorphous **Price (2002):** ≈ 1.10 €/kg
Characteristics:	stiffness, brittleness, good dimensional stability, inexpensive, smooth surfaces, susceptibility to stress cracking, limited chemical resistance
Identified by:	transparency, smell of styrene when burned, yellow flame, soot forming

Properties:
ρ = 1.05 g/cm^3
E = 3100– 3300 MPa
σ_B = 30–55 MPa
ε_s = –
ε_B = 1.5–3%
T_g = 90–100 °C
Thermal limits:
Short-time: ~ 90 °C
Long-time: ~ 80 °C

Structure:

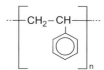

Often blended with:
PE, PP, PA, PMMA

σ - ε - Diagram:

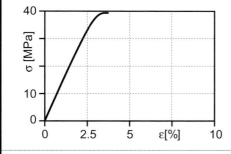

E-Modulus f (Temp):

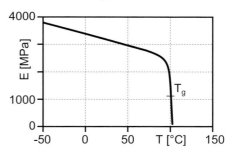

DSC - Diagram:

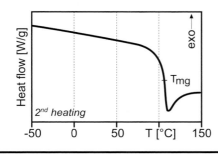

Coeff. of linear therm. expansion f (Temp):

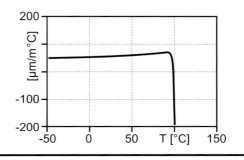

PS-I – Styrene Butadiene Copolymer (SB)

Type: amorphous, blend   **Price (2002):** ≈ 1.20 €/kg

Charcteristics: toughness, good dimensional stability, inexpensive,
                       sensitive to aging and environmental conditions

Identified by: opaqueness, smell of styrene when burned, bright flame, soot forming,
                       white failure surface

Properties*:

ρ = 1.00–1.05 g/cm^3
E = 1100–2800 MPa
σ_s = 15–45 MPa
ε_s = 1.1–6%
ε_B = 10–> 50%
T_g = ~ -85°C/100 °C

Thermal limits:
Short-time: ~ 60–80 °C
Long-time: ~ 50–70 °C
dependent on butadiene concentration

Structure:

Matrix:

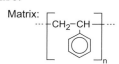

Rubber

Often blended with:
PE, PP, PA, PMMA

σ - ε - Diagram:

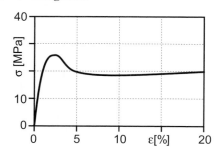

E-Modulus f (Temp):

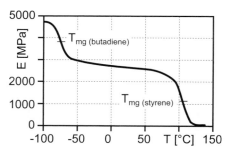

DSC - Diagram:

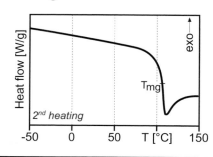

Coeff. of linear therm. expansion f (Temp):

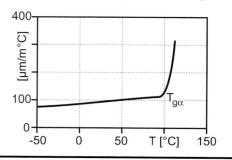

PS-S – Syndiotactic Polystyrene	

Type: semicrystalline (~35% when highly isotactic) **Price (2002):** ≈ 3.50 €/kg

Characteristics: brittleness, good thermal, chemical, and hydrolytic resistance, good stress crack resistance, good dimensional stability at equivalent amorphous and crystalline densities

Identified by: whitish, smell of styrene when burned, yellow flame, soot forming

Properties:
ρ = 1.28 g/cm³
E = 3400–3600 MPa
σ_s = 50–55 MPa
ε_s = –
ε_B = 1.5–2%
T_g = 90–100 °C
T_{pm}= 270 °C
Thermal limits:
Short-time: ~ 220 °C
Long-time: ~ 150 °C

Structure:

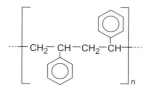

σ - ε - Diagram:

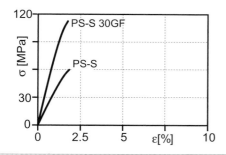

E-Modulus f (Temp):

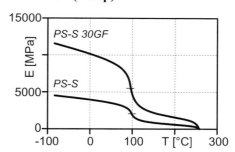

DSC - Diagram:

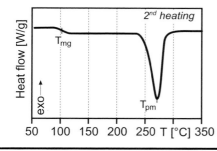

Coeff. of linear therm. expansion f (Temp):

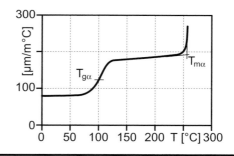

PSU – Polysulfone

Type: amorphous   **Price (2002):** ≈ 12.00 €/kg

Characteristics: high thermal stability, high dimensional stability, susceptibility to stress cracking, brittleness, often reinforced

Identified by: transparency, smell of styrene when burned, yellow flame, soot forming, difficult to ignite

Properties:

ρ = 1.24–1.25 g/cm^3
E = 2500–2700 MPa
σ_s = 70–80 MPa
ε_s = 5.5–6%
ε_B = 20–> 50%
T_g = 185–190 °C

Thermal limits:

Short-time: ~ 170 °C
Long-time: ~ 150 °C

Structure:

Often blended with:

ABS

σ - ε - Diagram:

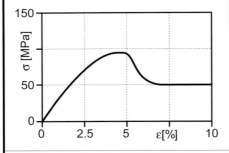

E-Modulus f (Temp):

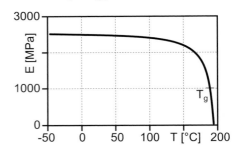

DSC - Diagram:

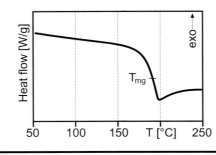

Coeff. of linear therm. expansion f (Temp):

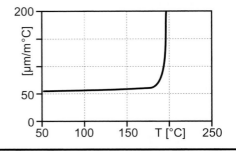

PTFE – Polytetrafluoroethylene

Type: semicrystalline (55–90%)   **Price (2002):** ≈ 20.00 €/kg

Characteristics: very good chemical resistance, high thermal stability,
good stress crack resistance, good environmental resistance,
high specific weight, tendency to creep, limited processability,
relatively high cost

Identified by: milky white color, blue-green flame, self extinguishing

Properties:

ρ = 2.13–2.23 g/cm^3

E = 400–750 MPa

σ_s = 20–40 MPa

ε_B = > 50%

T_g = 125–130 °C

T_{pm}= 325–330 °C

Thermal limits:

Short-time: ~ 280 °C

Long-time: ~ 240 °C

Structure:

$$\cdots\!\!-\!\!\begin{bmatrix} \overset{\displaystyle F}{\underset{\displaystyle F}{|}}\; \overset{\displaystyle F}{\underset{\displaystyle F}{|}} \\ C\!-\!C \\ \end{bmatrix}_n\!\!-\!\cdots$$

σ - ε - Diagram:

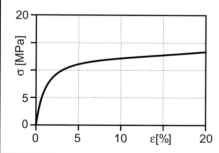

E-Modulus f (Temp):

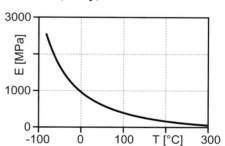

DSC - Diagram:

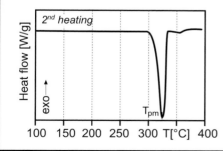

Coeff. of linear therm. expansion f (Temp):

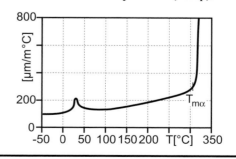

PVC-P – Polyvinyl Chloride (plasticized)

Type: amorphous   **Price (2002):** ≈ 1.50 €/kg

Characteristics: good environmental resistance, brittle at low temperatures,
                     self extinguishing poisonous fumes when burned

Identified by: transparency to opaqueness, smell of hydrochloric acid when burned,
                     difficult to ignite

Properties: ρ = 1.16–1.35 g/cm^3 E = 25–1600 MPa σ_B = 8–25 MPa ε_B = 170–400% T_g = -50–80 °C **Thermal limits:** Short-time: ~ 55–65 °C Long-time: ~ 50–55 °C *(large variability depending on the plasticizer, caution with T_g)*	**Structure:** $$\left[CH_2 - \underset{\underset{Cl}{\mid}}{CH} \right]_n$$ **Often blended with:** ABS, NBR, PE-C, PMMA

σ - ε - Diagram:

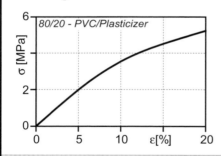

E-Modulus f (Temp):

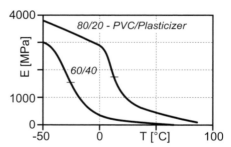

DSC - Diagram:

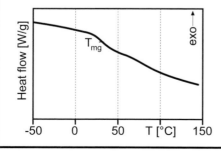

Coeff. of linear therm. expansion f (Temp):

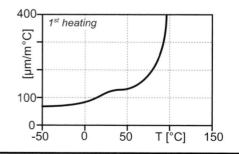

PVC-U – Polyvinyl Chloride (unplasticized)

Type:	amorphous	**Price (2002):** ≈ 0.80 €/kg

Characteristics: good dimensional stability, chemical and environmental resistance, poor toughness at low temperature, inexpensive, poisonous fumes when burned

Identified by: transparency to opaqueness, smell of hydrochloric acid when burned, self extinguishing

Properties:

ρ = 1.38–1.55 g/cm^3
E = 2700–3000 /mm^2
σ_s = 50–60 MPa
ε_s = 4–6%
ε_B = 10–50%
T_g = 80 °C

Thermal limits:
Short-time: ~ 70 °C
Long-time: ~ 60 °C

Structure:

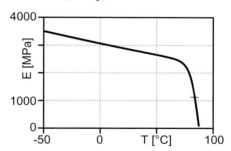

Often blended with:
ABS, NBR, PE, PMMA

σ - ε - Diagram:

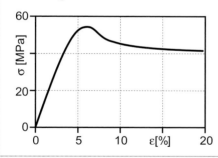

E-Modulus f (Temp):

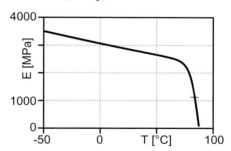

DSC - Diagram:

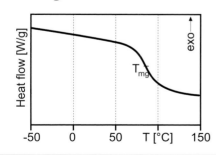

Coeff. of linear therm. expansion f (Temp):

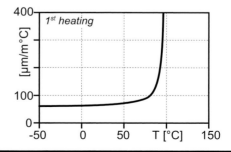

PVDF – Polyvinylidene Fluoride

Type:	amorphous, semicrystalline	**Price (2002):** ≈ 24.00 €/kg

Characteristics: good strength, high chemical and radaiation stability, high susceptibility to residual stress

Identified by: pungent smell when burned, difficult to ignite

Properties*:
ρ = 1.76–1.78 g/cm^3
E   = 2000–2900 MPa
σ_s = 50–60 MPa
ε_s = 7–10%
ε_B = 20–> 50%
T_g = 40 °C
T_{pm}= 170–175 °C
Thermal limits:
Short-time: ~ 150 °C
Long-time: ~ 120 °C

Structure:

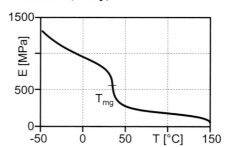

σ - ε - Diagram:

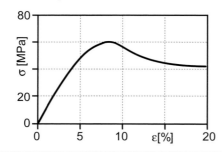

E-Modulus f (Temp):

DSC - Diagram:

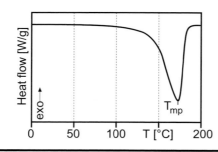

Coeff. of linear therm. expansion f (Temp):

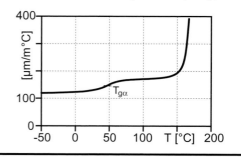

SAN – Styrene-Acrylonitrile Copolymer

Type: amorphous   **Price (2002):** ≈ 1.60 €/kg

Characteristics: high strength and stiffness, good dimensional stability, good chemical resistance, lower resistance against stress cracking than PS

Identified by: yellowish, transparency, smell of styrene when burned, bright flame, soot forming

Properties:

ρ = 1.08 g/cm^3
E = 3500–3700 /mm^2
σ_B = 65–85 MPa
ε_s = –
ε_B = 2.5–5%
T_g = 95–105 °C

Thermal limits:
Short-time: ~ 95 °C
Long-time: ~ 85 °C

Structure:

$$-\!\!\left[\,CH_2-CH-CH_2-\underset{\underset{CN}{|}}{CH}\,\right]_n\!\!-$$

σ - ε - Diagram:

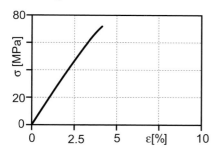

E-Modulus f (Temp):

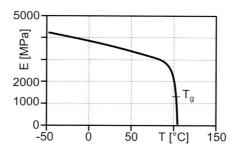

DSC - Diagram:

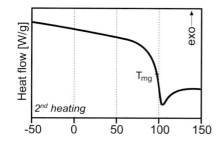

Coeff. of linear therm. expansion f (Temp):

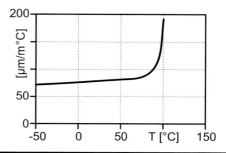

UP – Unsaturated Polyester Resin

Type: amorphous, crosslinked   **Price (2002):** ≈ 2.00 €/kg

Characteristics: resistance to aging, good chemical resistance, inexpensive,
                      good thermal stability, toughness to brittleness, often reinforced
                      with glass fibers, inexpensive

Identified by: transparency, smell of styrene when burned, does not melt

Properties*:

ρ = 1.2 g/cm^3
E = 3200–3500 MPa
σ_B = 50–77 MPa
ε_B = 1.2–2.5%
T_g = 70–150 °C

Thermal limits:

large variation depending on type
Short-time: ~ 180 °C
Long-time: ~ 100 °C

*for nonreinforced material

Structure:

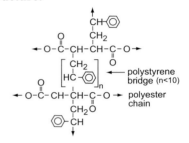

σ - ε - Diagram:

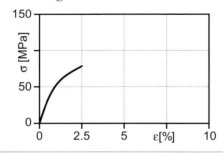

E-Modulus f (Temp):

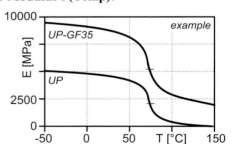

DSC - Diagram:

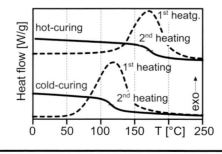

Coeff. of linear therm. expansion f (Temp):

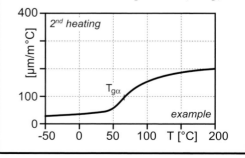

1   Subject Index

A

B

C

N

O

P

The Authors

Prof. Dr.-Ing. Dr. h.c. Gottfried Wilhelm Ehrenstein

Born 1937 in Danzig. He studied Mechanical Engineering at the Technical University in Hanover and earned his Ph.D. under the advisement of Prof. Matting. During the following 10 years he held various positions within the technical plastics application department of BASF, while at the same time serving as an educational representative. He qualified for professorship in 1976 and took a position as a faculty lecturer for Mechanical Engineering under the guidance of Prof. Macherauch at the Technical University of Karlsruhe. From 1977 to 1989, he was the Director for Plastics Material Technology at the University of Kassel. Since 1989, he is a professor at the University of Erlangen–Nuremberg. In addition, he served as the director of the Southern German Plastics Center (Süddeutsches Kunststoff–Zentrum, SKZ) in Wurzburg from 1987 to 1992. In 1992, he was awarded an honorary professorship at the Chemical Institute of the University of Quingdao, China. He was also awarded an honorary doctorate from the Technical University in Budapest, Hungary in 1996.

Dipl.-Ing. Gabriela Riedel

Born 1965 in Dux. She finshed her professional education as a laboratory technician for physical testing in 1986 at the Southern German Plastics Center. From 1986 to 1991, she studied Plastics Technology at the University of Applied Sciences Wurzburg–Schweinfurt. In 1991, she became supervisor at the Department for Chemical and Physical Testing, a part of the Department of Plastics Technology, at the University of Erlangen–Nuremberg.

Laboratory Technician Pia Trawiel

Born 1972 in Grimma. She finished her professional education as a laboratory technician for chemistry in 1986 at the VEB Chemie-anlagenbau Leipzig/Grimma. Since 1990, she has been employed at the Department of Chemical and Physical Testing, a part of the Department of Plastics Technology, at the University of Erlangen–Nuremberg, where she holds a position as deputy supervisor.

BURLINGTON COUNTY COLLEGE
COUNTY ROUTE 530
PEMBERTON NJ 08068